Solving DC and AC Circuits by Example Using MATLAB

Richard E. Haskell
Darrin M. Hanna

Oakland University, Rochester, Michigan

LBE Books
Rochester Hills, MI

ISBN 978-0-9824970-0-5

Published by LBE Books, LLC
1202 Walton Blvd.
Suite 214
Rochester Hills, MI 48307

www.lbebooks.com

Preface

This book uses Matlab as an aid to learning and understanding basic circuit analysis. Most introductory texts on circuit theory introduce Ohm's law and Kirchhoff's law very quickly and are then off to the races. In this book we spend a fair amount of time in Chapter 1 putting these laws and other basic concepts in a historical perspective. This will provide you with the knowledge of where the basic ideas of electrical science come from.

Chapter 2 discusses circuit elements including resistors, capacitors, and inductors. Ohm's law is covered in Chapter 2 and Kirchhoff's law's are described in Chapter 3. Sinusoidal signals and phasors are introduced in Chapter 4 where the concept of RMS values are described. This allows both DC and steady-state AC circuits to be solved in subsequent chapters. Examples involving voltage dividers, current division, and source transformation are given in Chapter 5. Circuits that include an ideal operational amplifier (op amp) are described in Chapter 6. Mesh and nodal analysis are covered in Chapter 7, while Chapter 8 covers superposition, Thevenin's theorem, Norton's theorem, and maximum power transfer. Average power and transformers are covered in Chapter 9 and the theory behind the operation of DC motors is given in Chapter 10. Basic material on vectors and matrices are included in the Appedices. By restricting the analysis of AC circuits to sinusoidal signals we do not require any knowledge of differential equations and we leave transient analysis to a later course.

All circuit problems in this book contain only independent voltage and current sources. All worked examples in the book show the calculations using Matlab. In most cases Matlab is used as a calculator and a good scientific calculator could be used to solve most of the examples. The use of Matlab clarifies all of the steps. However, we also use Matlab to create a variety of plots that will help you understand the material. PSpice is another computer-based tool that the electrical engineering student should learn. However, PSpice is a simulator that simulates the behavior of an electrical circuit. In PSpice you basically draw a schematic diagram of the circuit and push a button. It will show you the voltages and currents in the circuit and plot useful graphs. It is particularly useful in more complex circuits involving transistors and other non-linear elements. We don't cover such circuits in this book and therefore the use of Matlab is preferred where you need to understand how to solve the problem before using Matlab to calculate numerical results. We show how to use Matlab to plot PSpice-like graphs.

Many colleagues and students have influenced the development of this book. Their stimulating discussions, probing questions, and critical comments are greatly appreciated. Special thanks go to Michael Polis and Wayne Morrell who have provided important contributions to many of the examples and topics described in this book.

Richard E. Haskell
Darrin M. Hanna
Oakland University
Rochester, Michigan 48309

Solving DC and AC Circuits by Example Using MATLAB

Table of Contents

Chapter 1

History and Basic Concepts

1.1 Early History – Charge and Voltage

Imagine living without electricity. The Northeast blackouts of 1965 and 2003 gave some indications of how dependent we have become on electricity in our daily lives. But for most of recorded history, people lived their lives without electricity! When the English landed at Jamestown, Virginia in 1607 and the pilgrims landed in Massachusetts in 1620, no one in the world was using electricity. **William Gilbert**, an English physician to Elizabeth I and James I, had just published his book, *On the Magnet and Magnetic Bodies, and on the Great Magnet the Earth*, in 1600. In this book he describes in great detail the properties of *loadstones*, a variety of magnetite known to the ancient Greeks, Egyptians, and Chinese. Loadstones behave as magnets that have poles and attract iron. Gilbert showed that the earth behaved like a large magnet. Mariners had been using magnetic compasses as an aid to navigation for centuries. While most of his book is about magnets, Gilbert also describes the properties of *amber*, a hard, translucent, yellow substance also known to the ancient Greeks. When this substance is rubbed, it is able to pick up feathers and other light substances. Gilbert showed that there were many other substances that had this property of attraction including jet, a hard, black fossil substance, as well as diamond, sapphire, opal, amethyst, and many types of glass. Gilbert called such substances *electrics*, from *electron*, the Greek word for amber. He coined the word electricity to describe this attraction phenomenon. However, he made a clear distinction between the attraction of magnets and the attraction of his electrics. It would be over 250 years before the close interrelationship between electricity and magnetism would be established.

By the time that **Benjamin Franklin** was born in Boston, Massachusetts in 1706, **Francis Hauksbee** had demonstrated a primitive light source generated by rotating an evacuated glass globe containing mercury and holding his hand on the glass. The first such friction machine was made by **Otto von Guericke** around 1663 and consisted of a ball of sulphur that was rotated by a crank. The friction of hands held against this rotating sulphur ball would cause the ball to become charged. Similar machines were used for the next century to generate electrical charges that would then produce sparks and shocks for entertainment. A typical demonstration would have the performer charge a glass rod and then draw sparks from the hands and feet of a small boy hanging by silk cords. Benjamin Franklin, a Philadelphia printer at the time, attended such a show in Boston in 1743 and as a result became very interested in this electrical phenomenon. In 1747 he ordered a glass tube and other materials from **Peter Collinson** in England and carried out a series of electrical experiments over the next few years. He reported the results of his experiments in letters to Peter Collinson who had them printed all over

Europe. Franklin introduced a single-fluid theory of electricity and showed that the generation of what he called positive electricity was accompanied by the generation of an equal quantity of negative electricity. We now know that negative charges come from electrons and that positive charges come from protons; but the electron wouldn't be discovered for another 150 years. Franklin's famous kite experiment showed that lightning was a form of electricity and his invention of the lightning rod made him world famous.

Up until this time, the electrical energy generated by friction machines couldn't be stored. In 1746, **Pieter van Musschenbroek**, a Dutch scientist at the University of Leyden, invented what became known as the Leyden jar that could store electrical charges. (Ewald von Kleist, a German scientist, independently constructed a similar device in late 1745). It was a glass jar with metal foil wrapped on the outside and filled with water or lead on the inside that could be charged through a wire connected to a metal ball. The Leyden jar could be discharged, producing a spark, by connecting the metal ball to the outside foil. By 1749, Benjamin Franklin was using a Leyden jar to store electrical charges. He found that the outside foil contained an equal and opposite charge from the inside. This was the first *capacitor*, a device for storing electrical charge.

The symbol for a capacitor is shown in Fig. 1.1. Franklin realized that when objects were electrified (charged) with his two kinds of electricity, positive and negative, objects of the same sign repulsed each other while objects of opposite signs attracted each other. In 1755, he found that if he charged a can and then lowered a cork-ball into it there was no force on the cork ball. He suggested to his friend **Joseph Priestley** that "*possibly the mutual repulsion of the inner opposite sides of the electrised cann, may prevent the accumulating an electric atmosphere upon them and occasion it to stand chiefly on the outside.*"[1] Priestley repeated his experiment and suggested that this meant that the attraction (and repulsion) of charged bodies followed an inverse square law. Although Priestley published this results in 1767, it would be 18 years before **Charles Coulomb**, who was working in Paris, used his torsion balance to measure the forces on electrified bodies, and proved the result, now known as Coulomb's law, that the magnitude of the force between two charges, $q1$ and $q2$, separated by a distance r is given by

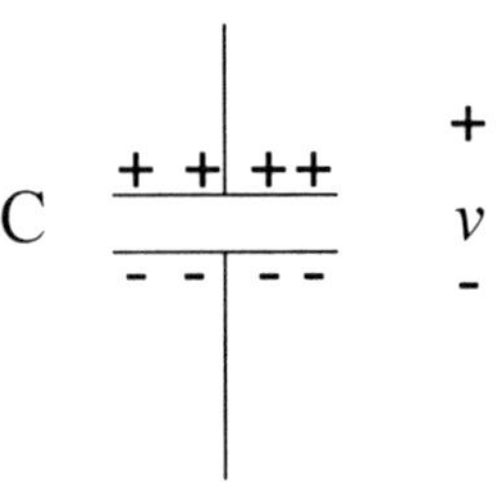

Figure 1.1
Symbol for a capacitor

$$F = \frac{1}{4\pi\varepsilon_0}\frac{q_1 q_2}{r^2} \tag{1.1}$$

where ε_0 is the permittivity of free space and the constant $1/4\pi\varepsilon_0$ has the value $10^{-7}c^2 = 8.99\times10^9$ Newton-meter^2/coulomb^2. The constant c is the speed of light, which is approximately 3×10^8 m/sec. The direction of the force is along the line joining the two charges. Like charges repel and opposite charges attract. The unit of charge is the *coulomb* (C), named after Charles Coulomb, and the electron charge is 1.602×10^{-16} coulombs.

---

[1] Robert S. Elliott, Electromagnetics, McGraw-Hill, New York, 1966, page 99.

It turns out that **John Robison** conducted similar experiments in Edinburgh in 1769 to discover the law of force between electric charges but this work wasn't discovered until after his death in 1822. **Henry Cavendish** also conducted definitive experiments in 1773 which demonstrated the inverse square law. However, like Robison, he did not publicize this work and it was not discovered for nearly a century. Thus, Coulomb, who did his work in 1785, gets the credit for discovering the inverse square law and the unit of charge is a *coulomb* and not a *robison* or a *cavendish*. This demonstrates the importance of quickly publishing new results, since otherwise there is always the risk that someone else will get the credit for your work.

Now consider the capacitor shown in Fig. 1.1 whose symbol looks like a pair of parallel plates. The top plate containing the positive charges could represent a metal foil wrapped on the inside of a Leyden jar while the bottom plate containing negative charges could represent the metal foil wrapped on the outside of the same Leyden jar. It took work, or energy, in the form of rubbing a glass rod to charge up the Leyden jar. The charges shown in Fig. 1.1 now have potential energy that can be used, for example, to move other charges.

Suppose that a small positive test charge is placed between the two plates in Fig. 1.1. It will experience a repulsive force from the top plate and an attractive force from the bottom plate. Thus, the net force will be downward and could be computed by adding the forces from all of the charges using Eq. (1.1). If we compute this net force and divide it by the charge of our test charge we get what is called the *electric field*. The electric field is an abstract concept that makes it easier to think about the effects that the charges on the plates will have on other charges. We talk about the force on our test charge due to the electric field where the electric field is created by the charges on the plates. We come to think of these electric fields as real and even draw pictures of them. However, they just represent the force that would be experienced by a unit charge at a particular point in space. In general, electric fields are a function of both position and time. Because electric fields are proportional to force, they have both a magnitude and direction, and are therefore *vectors*. We denote the vector electric field as **E**. Vectors are described in Appendix A.

Now back to our little test charge. We know that force times distance is work. The work per unit charge done by all of the other charges (or by the electric field) in moving our test charge from the top plate to the bottom plate is called the *voltage drop*, *v*, from the top plate to the bottom plate. The *voltage rise*, *-v*, from the bottom plate to the top plate is the work per unit charge done by an external force in moving our test charge from the bottom plate to the top plate. If our test charge was negative, then the voltage drop would be negative and the voltage rise will be positive.

The voltage *v* in Fig. 1 will be proportional to the total charge *q* on the top plate. (A charge of *-q* will always be on the bottom plate.) We could also say that *q* is proportional to *v* and write the equation

$$q = Cv \qquad\qquad (1.2)$$

where the constant of proportionality *C* is called the *capacitance*. The unit of voltage is the *volt* and the unit of capacitance is the *farad*. From Eq. (1.2) we see that a farad is a coulomb/volt. The capacitance of a capacitor, or a Leyen jar, depends only on the geometry of the device and the characteristics of the material between the plates.

If we connect a wire from the top plate to the bottom plate in Fig. 1.1, the electrons on the bottom plate will rush up the wire and combine with the positive charges to form neutral atoms. Thus, the capacitor will discharge, there will be no charges left on the plate, and the voltage v will be zero. This is what happened when Benjamin Franklin got a shock and produced sparks when he would touch a wire from the inside to the outside of his Leyden jar.

In this book we will use the SI system of units and the SI prefixes are shown in Table 1.1.

Table 1.1 SI Prefixes

Multiplier	Pefix	Symbol
10^{12}	terra	T
10^{9}	giga	G
10^{6}	mega	M
10^{3}	kilo	k
10^{-1}	deci	d
10^{-3}	mili	m
10^{-6}	micro	μ
10^{-9}	nano	n
10^{-12}	pico	p

When Leyden jars were charged using friction machines a large amount of charge and therefore large voltages could be achieved. For example, a typical Leyden jar might have a capacitance of 700 pF and a voltage of 175 kV. The total charge q stored would then be

$$q = Cv$$
$$= 700 \times 10^{-12} \times 175 \times 10^{3}$$
$$= 1.225 \times 10^{-4} \text{ coulombs}$$

This would correspond to a total number of electrons equal to

$$\text{No. of electrons} = 1.225 \times 10^{-4} \text{ coulombs}/1.6 \times 10^{-19} \text{ coul/electron}$$
$$= 7.66 \times 10^{14} \text{ electrons}$$

We will discuss capacitors in more detail in Chapter 7.

We define the voltage v to be the change in energy (work) dw of a small charge dq as it passes through the circuit. Thus,

$$v = \frac{dw}{dq} \qquad (1.3)$$

As we have seen, the problem with using a Leyden jar as a voltage source is that as soon as you use it and remove the charges the voltage goes away. The search was on for a continuous source of voltage that would provide a continuous source of charges.

Luigi Galvani was an Italian physician and anatomy professor who became famous in the 1780's for getting the muscles of dissected frog legs to twitch when he joined two wires of different metals that were connected to different parts of the frog leg. He attributed this behavior to *animal electricity* within the frog's muscle. However, the Italian physicist **Alessandro Volta** showed that animal electricity had nothing to do with it by replacing the frog's muscle with solutions of salts or acids and was able to obtain strong electrical effects. In 1800 he described his invention of the first chemical battery (called a *voltaic pile*) that generated current from the chemical reaction of zinc and copper discs separated from each other with cardboard discs soaked in a salt solution. This was a major discovery because it allowed a continuous current (moving charges) to be observed in a circuit. The unit of voltage, *volt*, is named after Alessandro Volta. A *volt* is the energy in *joules* required to move a charge of one *coulomb* through an element.

It was Benjamin Franklin who coined the word *battery* for a series of charged glass plates that he lined up and wired together. Following his experiments with electricity, Benjamin Franklin became a politician and diplomat, signed the Declaration of Independence in 1776, spent several years in France negotiating a peace treaty with England to end the revolutionary war, and returned to Philadelphia to take part in the constitutional convention of 1787. Part of Franklin's success as a diplomat was due to the interest his electrical experiments generated in European society. In 1790, a year after George Washington became the first president of the United States, Benjamin Franklin died. The following year **Michael Faraday** was born.

1.2 Current – Oersted, Ampere, Faraday, and Henry

The battery invented by Volta provided a continuous supply of charges that could pass through a wire in the form of a current. The current through a wire, i, is defined as the rate at which charges flow through wire. If an amount of charge, dq, passes a certain point in the wire in a time dt, the current at that point is given by

$$i = \frac{dq}{dt} \qquad (1.4)$$

The current will be positive if a net amount of positive charge passes the point in the direction of current flow. Today we know that it is really the electrons (which have a negative charge) that actually do the moving in the wire. Nevertheless, we stick with the convention that positive current is in the direction of motion of positive charges. Thus, in a real circuit the electrons are actually moving in the opposite direction of what we call the current.

Hans Christian Oersted was a Danish physicist and chemist who made an important discovery in 1820 at the University of Copenhagen. He had noticed that a compass needle fluctuated erratically during a thunderstorm and thought that there must be some relationship between electricity and magnetism. During one of his lectures, he had a long straight wire connected to a battery so that a current flowed through the wire. He put the compass needle at right angles to the wire but observed no effect. After the lecture he thought to try putting the compass needle parallel to the wire and observed that the needle deflected significantly. He then got a larger battery and collected some colleagues as witnesses and repeated the experiment. He found that if he placed the needle about three-quarters of an inch below the wire, the needle would deflect about 45°. If he placed the needle above the wire, it would deflect in the opposite direction. He also found that the effect would still occur if he put materials such as glass, metals, wood, water, resin, stoneware, and stones between the wire and the needle.

Thus, Oersted showed that a current passing through a wire behaves like a magnet. As soon as the French physicist **André-Marie Ampère** learned of Oersted's discovery, he repeated the experiment and made an even more important discovery. He reasoned that if a current-carrying wire behaved like a magnet, then two current-carrying wires should behave like two magnets and either attract or repel each other. He immediately showed this to be the case. Two parallel wires carrying current in the *same* direction will *attract* each other, and two parallel wires carrying current in the *opposite* direction will *repel* each other. He also found that the force exerted on a current element in one wire by current flowing in the second wire was always normal to the current element. He assumed, incorrectly as it turned out, that the force between two current-carrying elements is directed along the line connecting the two elements – as it is for charges in Coulomb's law. However, this is not the case.

We saw earlier that it is sometimes convenient to think of the force between two static charges in terms of an *electric field*. One of the charges produces the electric field, and then we say that the force on the second charge is due to the electric field. (Of course, it is really due to the other charge.) In the same way it is even more convenient to think of the forces between two currents in terms of what we call a *magnetic field*.

We picture electric and magnetic fields by drawing lines. Electric field lines originate on electric charges, so the electric field due to a single positive charge looks like the spokes of a wheel with straight arrows shooting out in all directions from the charge. However, there are no such things as "magnetic charges" on which magnetic fields can originate. Magnetic fields are generated by moving charges, i.e. currents, and the magnetic field lines are circles that go around the currents. Magnetic field lines are always closed lines. At any point in space the magnetic field has a magnitude and direction and is therefore a vector field just like the electric field. The symbol for the magnetic field is **B** and it is sometimes referred to as the *magnetic flux density*. Fig. 1.2 shows a magnetic field line resulting from a current going into the page represented by the **X**. The direction of the magnetic field is given by the *right-hand rule*: Point the thumb of your right hand in the direction of the current (into the page), and your curled fingers will point in the direction of the magnetic field.

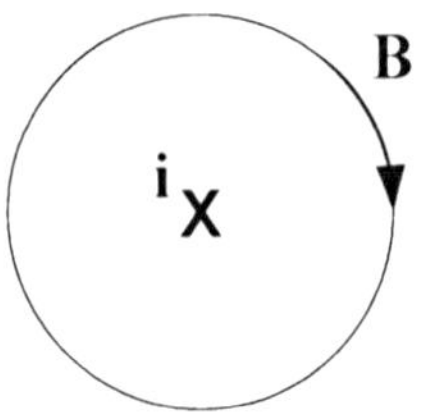

Figure 1.2
Magnetic field
generated by a current

Suppose now that in Fig. 1.2 we put another wire carrying current into the page parallel to the current **X** and located about where the **B** arrow is. Ampère discovered that the force $\Delta \mathbf{F}$ on a wire of length ΔL in a magnetic field **B** and carrying a current **i** is given by

$$\Delta \mathbf{F} = \left(\mathbf{i} \times \mathbf{B} \right) \Delta L \qquad\qquad (1.5)$$

where $\mathbf{i} \times \mathbf{B}$ is the cross product of the vectors $\mathbf{i}$ and $\mathbf{B}$. As shown in Appendix C, the cross product of two vectors is a third vector perpendicular to the first two in a direction given by the right-hand rule. In this case point the fingers of your right hand in the direction of $\mathbf{i}$, rotate them in the direction of $\mathbf{B}$, and your thumb will be pointing in the direction of the force. Using this rule you should be able to verify Ampère's observations that two parallel wires carrying current in the *same* direction *attract* each other, and two parallel wires carrying current in the *opposite* direction *repel* each other. We will use Eq. (1.5) in Chapter 10 to describe how a dc motor works.

A Note on Units

The unit of electric current is the ampere (A), named after André-Marie Ampère. In fact, in the International System of units (SI), electric current is taken as one of the base units from which all other units are derived. The other base units are shown in Table 1.2.

The official SI definition of current is that current which, if flowing in two parallel conductors of infinite length one meter apart would produce a force of 2×10^{-7} newtons (N) on each meter of length. The conductors must also be of negligible circular cross section and the whole thing must be in a vacuum. The meter, kilogram, and second have equally weird official definitions (look them up!).

All other electrical units can then be defined in terms of the ampere. For example, the coulomb is the amount of charge that a current of one ampere moves past a point in one second. We have already seen that a *volt* (V) is the energy in *joules* required to move a charge of one *coulomb* through an element; i.e., a volt is a joule/coulomb. A joule is a newton-meter, and a newton is a meter-kilogram/second^2. Thus, all electrical units can be expressed in terms of the SI base units, m, kg, s, and A. We often use the prefixes in Table 1.1 to indicate larger or smaller values of a particular unit. For example, a milliampere (mA) is 10^{-3} amperes and a kilovolt (kV) is 10^{3} volts.

Table 1.2 International System of Units (SI)

Quantity	Basic Unit	Symbol
Length	meter	m
Mass	kilogram	kg
Time	second	s
Electric current	ampere	A
Thermodynamic temperature	kelvin	K
Amount of substance	mole	mol
Luminous intensity	candels	cd

The First Electric Motor

At the age of 13 in 1804, **Michael Faraday** was hired as an errand boy for George Riebau's bookbindery in London. The following year he started a 7-year apprenticeship with Riebau to learn the craft of bookbinding. During these seven years, he became self-educated by reading many of the books that he was binding. He became most interested in science and dreamed of becoming a research scientist. But, due to his lack of education and humble beginnings, this goal seemed out of reach. He would attend scientific lectures and then write up what he heard in detail. In 1812 one of Riebau's customers, a prominent Londoner, made arrangements for Faraday to attend a lecture by **Sir Humphry Davy** at the Royal Institution. Davy was Professor of Chemistry at the Royal Institution and president of the Royal Society. He had become famous for his experiments (on himself) of the physiological effects of nitrous oxide, or laughing gas. Faraday attended the lecture and wrote up a detailed description of what he heard. His apprenticeship ended that year and he applied for a job at the Royal Institution, sending Davy a copy of his write-up of Davy's lecture. But Davy had no opening and Faraday started work at another bookbindery. However, in early 1813, a laboratory assistant at the Royal Institution was fired, and Faraday was hired to take his place. In the fall of 1813, Davy and his wife set sail for what would be an 18-month journey to France and Italy. They took Faraday with them as Davy's assistant and valet. While Davy's wife didn't let Faraday forget his position as a valet, Faraday was able to meet the top scientists on the continent including Ampère. When he returned to London from this voyage in 1815, he continued to work on a variety of experiments as Davy's assistant until 1821, when he was appointed Assistant Superintendent of the House of the Royal Institution.

In the fall of 1820, Davy informed Faraday of Oersted's discovery that an electric current moved a compass needle. After Ampère's demonstration of the force between two current-carrying wires, the race was on to produce rotary motion from electricity. In the spring of 1821, Davy and **William Wollaston** tried without success. In September of 1821, Faraday, who had read and duplicated the experiments of Oersted and Ampère, succeeded in getting a straight wire to continuously revolve about a permanent magnet. A schematic of his basic setup is shown in Fig. 1.3. A permanent magnet M immersed in a pool of mercury produces a magnetic field B as shown. A straight wire W is connected to a positive terminal and allowed to pivot about the point A. If a battery is connected between the + and − terminals, the mercury conducts the current flowing through the wire and the wire begins to rotate about the magnet. Using Eq. (1.5) can you tell which way the wire rotates about the magnet? (You need to keep reading to find out if your answer is correct).

This was the first demonstration of what would become electric motors. Faraday published this result in October 1821. However, he failed to reference the work of Davy and Wollaston and this strained his relationship with Davy. Faraday was elected as a fellow of the Royal Society in 1824 and was appointed Director of the Laboratory in 1825. Humphry Davy died in 1829.

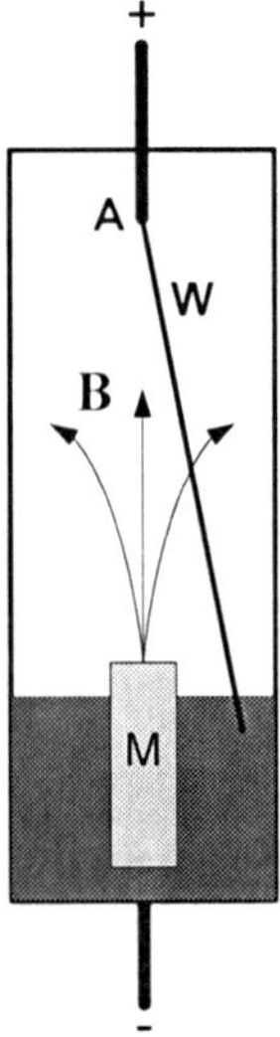

Figure 1.3
Faraday's
electromagnetic rotator

Faraday was interested in the basic science of electrical phenomena and would leave it to others to reduce the results to practical devices. The first electric motor to receive a patent was invented by an obscure blacksmith in Brandon, Vermont named **Thomas Davenport.**[2] Having heard that electromagnets were being used by the Penfield Iron Works near Crown Point, NY to extract iron from pulverized ore, Davenport traveled there and bought one of the electromagnets. He brought it back to Vermont, took it apart, and used it to make an electric motor that rotated continuously by using a brush and commutator. (See Chapter 10 for a discussion of how a brush and commutator are used in an electric motor.)

The electromagnets used to extract iron at Crown Point, NY were made by **Joseph Henry** who was teaching at the Albany Academy at the time. He had developed the world's largest electromagnets including one that could lift 3,600 pounds. Henry became a professor at Princeton University in 1833 and then became the first secretary of the Smithsonian Institution in Washington, DC where he served from 1845 until his death in 1878. (See Fig. 1.4.)

(a)
Statue of Henry in front on the
Smithsonian Castle in Washington DC

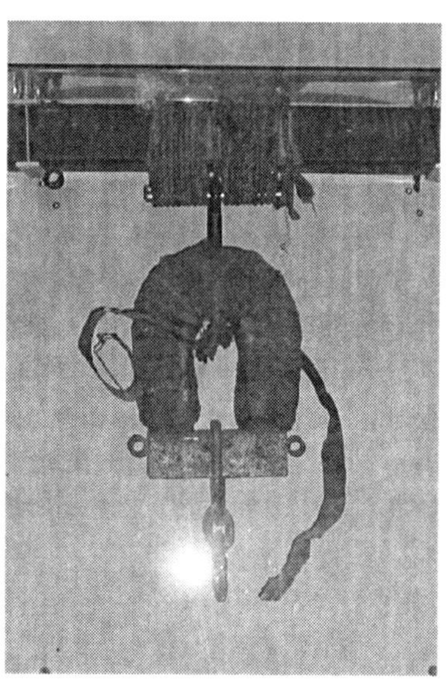

(b)
Electromagnet designed by
Henry on display in the National
Museum of American History

Figure 1.4

An electro-magnet is made by wrapping insulated wire around an iron core as shown in Fig. 1.5 and running current through the wire. Notice that if we apply the right-hand rule shown in Fig. 1.2 to determine the direction of the magnetic field created by the current, all of the magnetic field lines within the coil will point in the same direction and add up as shown in Fig. 1.5. Using an iron core, being a magnetic material, will greatly enhance the strength of the

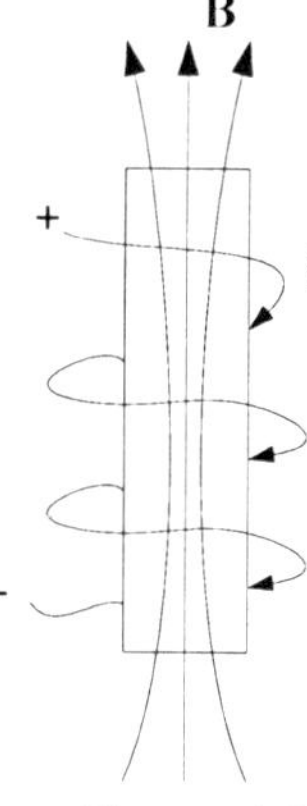

Figure 1.5
Making an electro-magnet

---

[2] Frank Wicks, "the blacksmith's motor," Mechanical Engineering, 1999.

magnetic field. In some of his early electro-magnets, Henry would wind 35 feet of wire around a horseshoe-shaped piece of iron to form about 400 turns. Later electro-magnets contained many more turns which increased the strength of the magnetic fields.

Another useful application of a coil of wire is to measure the amount of current in the wire by placing the coil between the poles of a permanent magnet and allowing the coil to rotate. The magnetic field produced by the current in the coil will try to align with the magnetic field of the permanent magnet and the amount of deflection of a needle connected to the coil will be a measure of the amount of current in the coil. Such a device is called a *galvanometer* and was used by many of the early experimenters to detect current in a circuit.

In 1835, Davenport demonstrated his working motor to Amos Eaton, who was Senior Professor at the Rensselaer Institute in Troy, NY. Eaton had co-founded the institute with Stephen Van Rensselaer as the Rensselaer School in 1824 making it the oldest engineering school in the country. Stephen Van Rensselaer paid Davenport $30.00 for the motor and gave it to the Rensselaer Institute. Thus, the oldest engineering school in the country had the first electric motor until it was apparently destroyed in a fire in 1862. In 1861 the school's name was changed to Rensselaer Polytechnic Institute (RPI). Before moving to Princeton, Joseph Henry served as an examiner at Rensselaer. Davenport had trouble getting a patent for his motor, but after meeting with Joseph Henry at Princeton and building new models he finally received the first U. S. patent for an electric motor, U.S. Patent No. 132, on Feb. 25, 1837.

Henry, like Faraday, was more interested in the science than in creating practical motors. In 1831, he had demonstrated an "Electro-magnetic Engine" that produced a continuous rocking motion of an electromagnetic bar magnet by a clever scheme of breaking the circuit in such a way as to alternately reverse the polarity of the magnet. However, Henry considered this device, as well as Davenport's electric motor to be a "philosophical toy" with no practical applications other than as interesting demonstrations for his students in his lectures.

Electromagnetic Induction

Once Oersted had discovered that a current could produce a magnetic field, there was an increasing feeling that the opposite might be true – that a magnetic field could produce a current. It was thought that if you placed a coil in a big enough magnetic field you might detect a current. Thus, Henry's big electromagnets were the equipment of choice in this quest. Both Faraday and Henry were looking for such currents in 1831. But for a long time, no matter how big the magnetic field was, they could detect no current in a coil that was placed in the magnetic field.

Faraday tried many arrangements with no success. Then in the fall of 1831, he wrapped 203 feet of copper wire around a large wooden block to create a coil such as that shown in Fig. 1.5. He wrapped a second 203 feet of similar wire around the same wooden block having each coil insulated from the other. He connected the first coil to a battery and the second coil to a galvanometer. When he connected the battery to the first coil, he noticed that the galvanometer suddenly twitched but then remained unmoved when the large current was flowing in the first coil. However, when he disconnected the battery, the galvanometer twitched again. He had discovered that in order to induce a current in the second coil the magnetic field generated by the first coil *must be changing*.

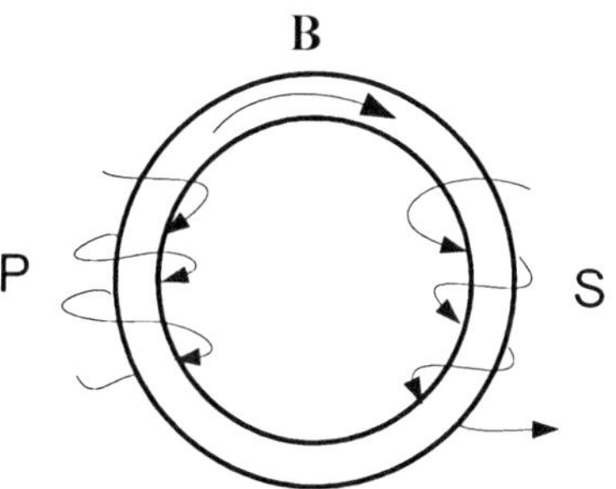

Figure 1.6
Setup for Faraday's demonstration of electromagnetic induction

To study the effect in more detail, he had a welded ring of soft iron made and wrapped about 75 feet of copper wire around about half of the ring to make a primary coil labeled P in Fig. 1.6. He then wrapped about 60 feet of copper wire around the other half of the ring to form the secondary coil S shown in Fig. 1.6. He connected the secondary coil to a galvanometer and when he connected the primary coil to a battery the reaction of the galvanometer was far greater than the previous experiment without the iron ring. But again, once the circuit was closed, the galvanometer had no reaction at all. The iron core produced a much larger magnetic flux within the secondary coil and the *change of this magnetic flux* is what induced the current in the secondary coil. This phenomenon of a current in a primary coil inducing a current in a secondary coil is called *mutual induction*. Faraday also showed that if the magnetic field of a permanent magnet was changed within a coil of wire, a current would be induced in the coil. Faraday reported the results of his electromagnetic induction experiments before the Royal Society on November 24, 1831.

Meanwhile in America, Joseph Henry would also demonstrate mutual induction by using his electro-magnets, but he would not publish his results until the summer of 1832. However, Henry is credited with discovering *self-induction*. He noticed that if he connected a battery to an electro-magnet of the type shown in Fig. 1.5 and then opened the switch a spark would be produced at the switch. A current is induced in the same coil due to the changing magnetic field (going to zero when the switch is opened).

In 1834, a professor of physics at the University of St. Petersburg in Russia named **Heinrich Lenz** came up with a law to determine the direction of the induced current due to electromagnetic induction. He stated the law in terms of visualizing how currents in magnetic fields would move as follows: *"Whenever a metallic conductor is moved in the vicinity of a galvanic current or of a magnet, a galvanic current will be induced, which has a direction such that it would have caused the wire, if it were at rest, to move in a direction opposite to the one given, provided that the wire were movable only in the direction of the motion and in the opposite direction."*[3] For example, consider the experimental results of Ampère that parallel wires carrying current in the same direction attract each other. If instead of running current in the second parallel wire we simply move it in the magnetic field toward the first wire then a current will be induced in this second wire that is opposite to the current in the first wire. As a second example, consider Faraday's electromagnetic rotator shown in Fig. 1.3. If current flows from top to bottom, the wire will rotate clockwise looking down on the rotator. If we now

---

[3] Morris H. Shamos, Ed., *Great Experiments in Physics – Firsthand Accounts from Galileo to Einstein*, originally published by Holt, Rinehart and Winston, New York, 1959; republished by Dover Publications, New York, 1987, pages 161-162.

disconnect the battery, replace it by a wire, and just grab the wire and rotate it in the same clockwise direction, a current will be induced that flows from bottom to top. We have just created a *generator* that generates electrical energy from mechanical energy!

In fact, Davenport's electric motor was a commercial failure being about 40 years ahead of its time. Before electric motors became in widespread use, falling water and steam were used to rotate the equivalent of Davenport's motor to generate electricity. Thus, Davenport, who died dejected in 1851 without fulfilling his dream of electric motors being used for transportation, would have made faster progress by rotating his machine mechanically to generate electricity! In 1910 the Vermont Electrical Association met in Brandon, VT and unveiled the plaque shown in Fig. 1.7 to commemorate the contribution of Thomas Davenport to the development of the electric motor. The inscription on the plaque reads

Figure 1.7
Thomas Davenport plaque in Forestdale, VT
just outside of Brandon, VT

A coil of wire of the type shown in Fig. 1.5 behaves like an *inductor* in an electric circuit. A time-varying electric current $i(t)$ will produce a time-varying magnetic flux $\phi(t)$ which, in turn, will induce a voltage $v(t)$ across the coil. This voltage is proportional to both the time rate of change of $\phi(t)$ and the number of turns in the coil. This is called Faraday's law. As a result, the voltage $v(t)$ across the coil is also proportional to the input current $i(t)$ and can be written as

$$v(t) = L \frac{di(t)}{dt} \tag{1.6}$$

The constant L is called the *inductance* and the unit of inductance is the *henry* named for Joseph Henry. The use of the symbol L for inductance is in honor of Heinrich Lenz.

Faraday was named Fullerian Professor of Chemistry at the Royal Institution in 1833. He continued generating a prodigious output of experimental results until his retirement in 1862. All through the 1830's and 1840's, Faraday tried to explain the results of his electromagnetic experiments in terms of invisible "lines of force" that today we use to help picture electric and magnetic fields. But to Faraday these lines of force were real and he pictured charges and currents producing these lines of force which existed in the space between charges and produced the forces felt by other charges and currents. He even described how light might be the result of somehow plucking these lines of force to produce a vibration. Faraday had no mathematical training and could not express his ideas mathematically to other scientists who still clung to the current action-at-a-distance philosophy. His ideas of a field theory to explain electromagnetism would have to wait until 1857 when a young fellow at Trinity College in Cambridge named **James Clerk Maxwell** sent him a copy of a research paper he had written entitled *"On Faraday's Lines of Force."* As we will see in Section 1.4 Maxwell wrote a follow-on paper in 1864 that would change forever our view of electromagnetism.

After retiring in 1862, Faraday left his second floor apartment at the Royal Institution, which he and his wife Sarah had occupied since 1813, and moved into a house at Hampton Court provided to him by Queen Victoria. He died there in 1867 and was buried in Highgate Cemetery with a private funeral.

When Joseph Henry died in 1878 while still serving as the first Secretary of the Smithsonian Institution, he had the largest funeral that Washington DC had seen since the death of Abraham Lincoln.

1.3 Electric Circuits – Ohm and Kirchhoff

In 1827 at the age of 38, **Georg Simon Ohm**, a German physics professor, wrote a book called *The Galvanic Circuit Investigated Mathematically*, in which he described what we now call Ohm's Law. This law states that the current, i, flowing through a conductor between two points is proportional to the voltage, v, across the two points and inversely proportional to the resistance, R, of the conductor between the two points. Mathematically, we can write

$$i = \frac{v}{R} \tag{1.7}$$

Here, and in the remainder of this Chapter, we assume the passive sign convention that will be discussed in Chapter 2.

In a circuit, we use circuit elements to model the electrical activity that is going on. The resistance of a conductor is represented by the symbol for a resistor shown in Fig. 1.8. We will discuss Ohm's law in more detail and show how to use it to solve circuit problems in Chapter 2.

The symbol for a battery and the more general symbols for dc and ac independent voltage sources are shown in Fig. 1.9.

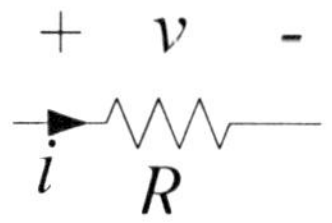

Figure 1.8
Symbol for a resistor

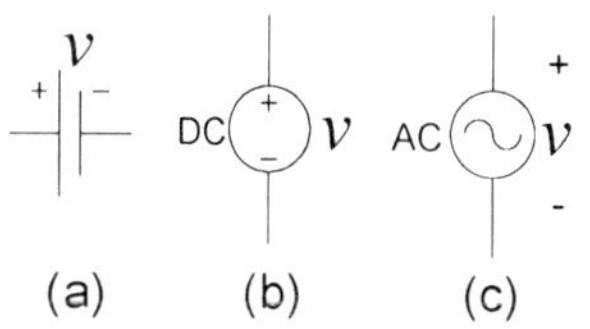

Figure 1.9
(a) Symbol for a battery
(b) Symbol for a dc independent voltage source
(c) Symbol for an ac independent voltage source

We will normally use the symbol shown in Fig. 1.9b when we want an independent dc voltage source. An independent voltage source maintains a constant voltage, v, independent of the amount of current supplied by the voltage source. This is an ideal that is only approximated for a real voltage source. A real battery, for example, has an internal resistance and the voltage supplied by such a battery will decrease if large currents are drawn from the battery. A better model of such a battery would be an independent voltage source in series with a resistor.

If we were to connect an independent voltage source across a resistor, we would get the circuit shown in Fig. 1.10. The symbol $\perp$ represents *ground* and is taken to be a reference voltage of zero. The voltage at point A is the voltage source v_0 and the voltage at point B is the voltage across the resistor v_R. The line between point A and B is a short circuit with a zero voltage drop. Thus, the voltage at point A is the same as the voltage at point B and $v_R = v_0$. In a real circuit we would connect points A and B with a wire. We assume that the wire has zero resistance so that it is a true short circuit. Real wires have small, but finite resistances that would be modeled by including a resistor symbol in the circuit.

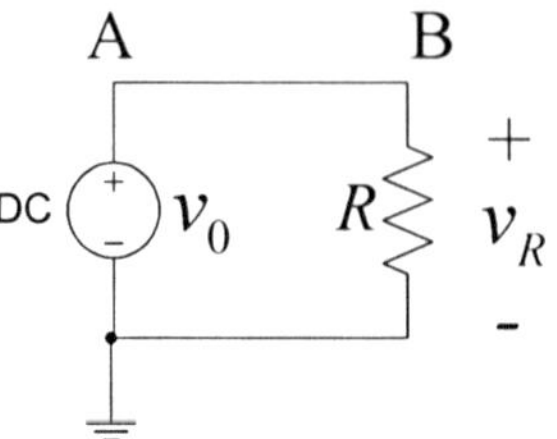

Figure 1.10
An electric circuit

In addition to an independent voltage source, a circuit might contain an independent current source whose symbol is shown in Fig. 1.11. An independent current source delivers a constant current i to the circuit independent of the voltage across the current. While an independent voltage source can be thought of as a battery, there is no corresponding device that is an ideal current source. Amplifier circuits can be designed to approximate an ideal current source.

Figure 1.11
An ideal current source

All of the circuits that we will analyze in this book will include only independent voltage and current sources. When analyzing transistor circuits, it is useful to model *dependent* voltage and current sources. The outputs of these dependent sources depend on some other current or voltage in the circuit. The symbols for these dependent voltage and current sources have a diamond shape. We will not consider dependent voltage and current sources in this book.

Kirchhoff's Current and Voltage Laws

Gustav Kirchhoff was also a German physicist who, while still a student at the Albertus University of Königsberg in 1845, formulated what have become known as Kirchhoff's Current Law (KCL) and Kirchhoff's Voltage Law (KVL). You will see that these laws are almost self-evident, but their application together with Ohm's law will allow us to solve any circuits problem.

Kirchhoff's current law (KCL) states that *the algebraic sum of currents entering (or leaving) a node must be zero*. This makes sense because otherwise charge would build up at the node and produce an extra voltage that is not part of the circuit. The directions of currents are assigned arbitrarily in a circuit by drawing arrows as shown in Fig. 1.12. Applying KCL to node A in that circuit (taking currents entering a node as positive) we can write

$$i_1 - i_2 - i_3 = 0 \qquad (1.8)$$

Kirchhoff's voltage law (KVL) states that *the algebraic sum of voltages around any loop must be zero.* This also makes sense because otherwise we would end up with a different voltage at the same point after going around the loop. In Fig. 1.12 we could move clockwise around the left loop, the right loop, or the entire outer loop. We will take the sign of each voltage as the polarity of the terminal first encountered in going around the loop. These polarities are arbitrary and indicated by using + and − signs. For example, applying KVL to the left loop in Fig. 1.12 we can write

$$-v_0 + v_1 + v_3 = 0 \qquad (1.9)$$

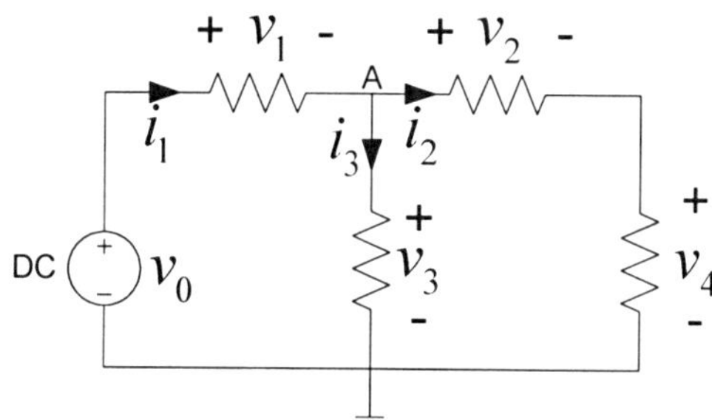

Figure 1.12
Circuit to illustrate Kirchhoff's Laws

Similarly, applying KVL to the right loop in Fig. 1.12 we can write

$$-v_3 + v_2 + v_4 = 0 \qquad (1.10)$$

Finally, applying KVL to the entire outer loop in Fig. 1.12 we can write

$$-v_0 + v_1 + v_2 + v_4 = 0 \qquad (1.11)$$

In addition to resistors, the two other passive circuit elements are capacitors and inductors. The circuit symbols for an inductor and a capacitor are shown in Fig. 1.13.

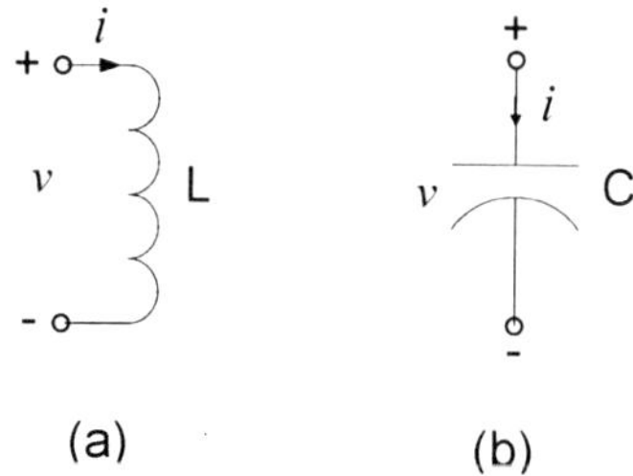

(a)   (b)

Figure 1.13
Circuit symbols for (a) an inductor and (b) a capacitor

From Eq. (1.6) we know that the voltage across the inductor is given by

$$v = L\frac{di}{dt} \tag{1.12}$$

The current through an inductor would then be given by

$$i = \frac{1}{L}\int v\,dt \tag{1.13}$$

From Eq. (1.2) we know that the charge on a capacitor is given by

$$q = Cv \tag{1.14}$$

and from Eq. (1.4) the definition of current is

$$i = \frac{dq}{dt} \tag{1.15}$$

Substituting Eq. (1.14) into Eq. (1.15) we see that the current through a capacitor is given by

$$i = C\frac{dv}{dt} \tag{1.16}$$

The voltage across a capacitor would then be given by

$$v = \frac{1}{C}\int i\,dt \tag{1.17}$$

Note that for dc circuits, the voltage across an inductor is zero and the current through a capacitor is zero. Thus, in a dc circuit an inductor looks like a short circuit and a capacitor looks like an open circuit. We only get a voltage across an inductor and a current through a capacitor in time-varying circuits. We will discuss capacitors and inductors in more detail in Chapter 3. We will show in Chapter 4 how circuits containing capacitors, inductors, resistors, and sinusoidal voltage and current sources can be analyzed by considering the impedances (corresponding to resistances in dc circuits) to be *complex numbers*.

1.4 Electric and Magnetic Fields – Maxwell, Hertz, and Einstein

James Clerk Maxwell was born in Scotland in 1831, the year that Faraday discovered electromagnetic induction. After being schooled at the Edinburgh Academy, he enrolled in Edinburgh University at the age of 16 and then moved to Cambridge in

1850 at the age of 19 and received his degree from Trinity College with high honors in 1854. He stayed at Trinity College for two years and then received a professorship at Marischal College in Aberdeen, Scotland in 1856. He was appointed Professor of Natural Philosophy at King's College, London in 1860.

While at Trinity College in Cambridge, he read all of Faraday's writings on electricity and magnetism, and in 1855, wrote his first of three major papers on the subject. This first paper took Faraday's ideas of lines of force and gave them a mathematical foundation. Maxwell used an analogy of an imaginary weightless and incompressible fluid that flowed through tubes that represented the lines of force. The equations he derived only worked for stationary lines of force, but it was a start and it correctly predicted the behavior of static electric and magnetic fields.

When Maxwell joined King's College in London in 1860 he returned to his study of electromagnetism with the goal of obtaining a mathematical representation of Faraday's lines of force that would explain all known electromagnetic behavior including electromagnetic induction. This time he changed his visual model from one of imaginary fluid-filled tubes to one of tiny imaginary close-packed spherical cells that could rotate and exert pressure on neighboring cells causing them to spin in the opposite direction. By means of this complicated model of spinning cells, Maxwell was able to derive equations that were able to explain what was then known about electromagnetism: 1) the attraction of unlike magnetic poles and repulsion of like magnetic poles (Gilbert), 2) the attraction of unlike charges and repulsion of like charges according to an inverse-square law (Coulomb), 3) the creation of a magnetic field by a current in a wire and the force on two current-carrying wires (Oersted and Ampère), and 4) the induced current in a wire due to a changing magnetic field (Faraday).

Maxwell published a paper in 1861 called *On Physical Lines of Force* that described this mechanical model of electromagnetism. However, in addition to explaining all known electromagnetic phenomena, this model predicted two new physical phenomena related to electromagnetism. The first was that a new kind of current that Maxwell called a *displacement current* would be produced every time an electric field changed. The second was that this displacement current would produce a magnetic field just like a normal conduction current and this changing magnetic field would produce a changing electric field, and on and on. This would lead to a wave-like behavior, and when Maxwell calculated what the speed of these waves would be, he discovered that the value was very close to the then known speed of light.

Maxwell was ecstatic and was sure that he was onto something big. He then abandoned his imaginary mechanical model of spinning cells in favor of a pure field theory of electromagnetism. In December of 1864 he introduced his most famous paper, *A Dynamical Theory of the Electromagnetic Field*, at a presentation to the Royal Society and he published it in 1865 in seven parts in the *Philosophical Transactions*. In this paper he presented what has become known as Maxwell's equations.

In terms of modern notation, Maxwell's equations are shown in Fig. 1.14. The first four equations are what we call Maxwell's equations and the last two equations are the constitutive equations that relate the electric flux density $\mathbf{D}$ to the electric field intensity $\mathbf{E}$ through the permittivity ε and the magnetic flux density $\mathbf{B}$ to the magnetic field intensity $\mathbf{H}$ through the permeability μ.

$$\nabla \cdot \mathbf{D} = \rho \quad (1) \qquad \nabla \cdot \mathbf{B} = 0 \quad (2)$$

$$\nabla \times \mathbf{E} = -\frac{\partial \mathbf{B}}{\partial t} \quad (3) \qquad \nabla \times \mathbf{H} = \mathbf{J} + \frac{\partial \mathbf{D}}{\partial t} \quad (4)$$

$$\mathbf{D} = \varepsilon \mathbf{E} \quad (5) \qquad \mathbf{B} = \mu \mathbf{H} \quad (6)$$

Figure 1.14 Maxwell's equations

The symbols used in these equations are vector calculus symbols that are probably not familiar to you, but the meanings of these equations are not difficult to understand. The first equation is a statement of Coulomb's law that says that the *divergence* of the electric flux density is equal to the charge density ρ. This means that the electric field lines (Faraday's lines of force) will extend radially from a charge as shown in Fig. 1.15. The force on a test charge located on the surface of the sphere shown in Fig. 1.15 will be proportional to the flux density $\mathbf{D}$; that is, it will be proportional to the number of lines that cross a unit area on the sphere. But the area of a sphere is $4\pi r^2$, so that the force must be inversely proportional to the square of the distance between the charges in accordance with Coulomb's law.

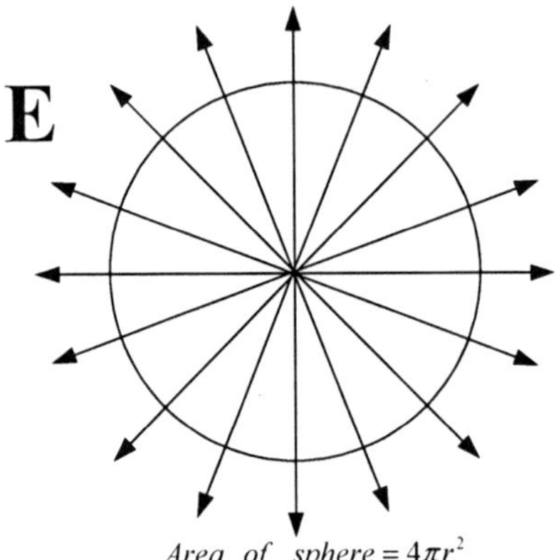

Figure 1.15 Electric field of a point charge

The second of Maxwell's equations in Fig. 1.14 states that the *divergence* of the magnetic flux density $\mathbf{B}$ is zero. This means there are no magnetic charges so that all magnetic field lines must be closed lines as shown by our example in Fig. 1.2.

Maxwell's equation number (3) in Fig. 1.14 is a statement of Faraday's law of electromagnetic induction and states that a changing magnetic field induces a *curl* of $\mathbf{E}$. You can think of the direction of $\mathbf{E}$ as sort of curling around the magnetic field lines. Another way of stating this equation is to say that the rate of change of magnetic flux density through an area A induces an electromotive force (voltage) equal to the line integral of $\mathbf{E}$ around the area A. That is, the voltage between two points is calculated by taking the component of the electric field at each point along a path and adding them up.

Finally, the last of Maxwell's equations, number (4) in Fig. 1.14, says that the total current, made up of the conduction current density $\mathbf{J}$ and the displacement current $\partial \mathbf{D}/\partial t$ will produce a *curl* of the magnetic field intensity $\mathbf{H}$. Without the displacement current $\partial \mathbf{D}/\partial t$ this is just a statement of Ampère's law. However, when Maxwell added the displacement current term it made all the difference in the world. For now in free space, when $\mathbf{J} = 0$, there is still a current source in Maxwell's equation number (4) – the displacement current $\partial \mathbf{D}/\partial t$. In this case, if Maxwell's equations (3) and (4) are combined, they produce a *wave equation* that predicts that transverse components of the electric and magnetic field vectors will propagate as electromagnetic waves at a speed given by $c = 1 / \sqrt{\mu_0 \varepsilon_0}$ where μ_0 is the permeability of free space and ε_0 is the permittivity of free space. When the values for these constants are plugged into the equation for c the speed of the electromagnetic waves turns out to be $3 \times 10^8 \, m/sec$, which is the speed of light!

When Maxwell published this groundbreaking paper, most other scientists were not convinced. They found the mathematics about fields difficult and they had long believed in action-at-a-distance and not in some mysterious activity going on in the space between charges and currents.

In addition to his work on electromagnetism, Maxwell made significant contributions in other areas including the theory of color vision, understanding the nature of Saturn's rings, and the kinetic theory of gases. However, his work on electromagnetism is his crowning achievement. As the American physicist Richard Feynman put it,

> "From a long view of the history of mankind - seen from, say, ten thousand years from now - there can be little doubt that the most significant event of the 19th century will be judged as Maxwell's discovery of the laws of electrodynamics. The American Civil War will pale into provincial insignificance in comparison with this important scientific event of the same decade".[4]

Maxwell retired from King's College in 1865 and returned to Glenlair, his family country estate in Scotland. While there, he did research and wrote up his work on electromagnetism. In 1871, Cambridge University persuaded him to return to Cambridge to set up a new laboratory for teaching and research. The new laboratory was completed in 1874 and named the Cavendish after Henry Cavendish, a reclusive 18th century chemist. Maxwell discovered unpublished works of Cavendish that showed that he had demonstrated the inverse square law of electric charges before Coulomb and had discovered Ohm's law 50 years before Ohm. In 1898 **J. J. Thomson** would discover the electron at the Cavendish laboratory.

Maxwell died in 1879 at the age of 48. At the time, most scientists were not convinced of his prediction of electromagnetic waves. They had never been observed. No one knew how to generate them or to detect them. Had he lived eight more years, he would have learned that **Heinrich Hertz** succeeded in generating and detecting his electromagnetic waves.

The Discovery of Electromagnetic Waves

Heinrich Hertz went to the University of Berlin in 1878 at the age of 21 to study under **Hermann von Helmholtz** and Gustav Kirchhoff. He received his doctorate in 1880 and continued working at Berlin until 1883 when he moved to the University of Kiel and continued to study Maxwell's work on electromagnetism. He became a

---

[4] Richard P. Feynman, Robert B. Leighton, and Matthew Sands, *The Feynman Lectures on Physics, Vol. II – Mainly Electromagnetism and Matter*, Addison-Wesley, Reading, MA, 1964, page 1-11.

professor of experimental physics at the University of Karlsruhe in 1885. Helmholtz had offered a prize for anyone who could provide an experimental test for Maxwell's theory. For a number of years, Hertz had tried to devise a scheme to detect either Maxwell's displacement current or electromagnetic waves.

In 1887, Hertz performed an experiment involving a block of paraffin between the plates of a capacitor, which convinced him that electromagnetic waves were being created when he observed sparks all over his apparatus as the capacitor was rapidly charged and discharged. He set about to do careful experiments that would show that electromagnetic waves propagated in free space. He generated the waves by producing sparks in a gap as a result of high voltages generated by an induction coil. He was able to detect electromagnetic waves that had traveled about 2 meters in free space by having the waves induce a current in a circular loop of copper wire that behaved like a resonant circuit and observing a tiny spark across a small spark gap in the wire.

Hertz was able to improve his apparatus and send electromagnetic waves up to 16 meters across the room. He was able to verify such wave properties as reflection, refraction, and polarization. He also measured standing waves with a wavelength of about 30 cm. He published accounts of these experiments in 1888 and 1889, and collected all of his papers in book form in 1893. In honor of Heinrich Hertz, the unit of frequency is the *hertz*, where a hertz is a cycle per second.

Hertz died in 1894 at the age of 37. It would therefore be up to a young Italian inventor named **Guglielmo Marconi**, starting in 1895, to show that "Hertzian waves" could be detected at long distances – first across his backyard and then in 1901 across the Altantic ocean! The age of *wireless telegraphy* had begun.

Electromagnetism and Relativity

In 1905 a 26-year-old patent clerk in Bern, Switzerland named **Albert Einstein** wrote a paper that would change our very concepts of space and time. Einstein was born the year that Maxwell died. Everyone has heard of Einstein's special theory of relativity that predicts seemingly unusual effects such as lengths of bodies contracting and time slowing down as bodies move at high speeds approaching the speed of light. What isn't as well known is that bodies don't need to be traveling near the speed of light to observe relativistic effects. What follows from Einstein's 1905 paper on special relativity is that the magnetic field is a purely relativistic effect. Thus, you observe relativity every day when you listen to the radio or watch TV!

The title of Einstein's paper on relativity is "*On the Electrodynamics of Moving Bodies*" and what bothered Einstein were inconsistencies in the interpretation of Maxwell's equations. The first sentence in the 1905 paper reads, "*It is well known that if we attempt to apply Maxwell's electro-dynamics, as conceived at the present time, to moving bodies, we are led to asymmetry which does not agree with observed phenomena.*"

Einstein then went on to contrast the different ways of explaining what happens when a magnet and conductor are in relative motion depending upon whether the magnet is moving or the conductor is moving. To get the basic idea, recall the force on a current element in a magnetic field from Eq. (1.5). We can rewrite this equation in terms of the force on a moving charge in a magnetic field as

$$\mathbf{F} = q\mathbf{v} \times \mathbf{B} \qquad (1.18)$$

which is called the *Lorenz force*. Note that only moving charges feel a force in a magnetic field. A stationary charge only feels a force due to an electric field given by

$$\mathbf{F} = q\mathbf{E} \qquad (1.19)$$

Now consider a current that is flowing in a conductor. It will produce a magnetic field like the one shown in Fig. 1.2. Suppose an electron with a negative charge is moving with a velocity $\mathbf{v}$ parallel to the conductor in the same direction as the electron flow in the conductor (i.e. opposite to the direction of the current). By Eq. (1.18), it will experience a force that will pull it in the direction of the conductor.

Now consider a coordinate system that is moving along with the electrons in the conductor at the same speed as these current electrons, and let the external electron that was moving parallel to the conductor have the same velocity as the current electrons in the conductor. That is, in the frame of reference that is moving with the current electrons, these electrons and the test electron on the outside are at rest. Now the fixed positive charges that were at rest in the original frame of reference will be moving in the new frame of reference and will produce a magnetic field. But the test electron is at rest in this moving frame of reference and therefore will experience no Lorenz force from Eq. (1.18). But clearly it must still move toward the conductor. It turns out that the relativistic contraction of the conductor moving at the current electron's speed is just enough to change the volume of the conductor so that the charge densities of the positive and negative charges are different. This will produce a net charge on the conductor that will produce an electric field in this moving frame of reference. The test electron that is stationary in this reference frame will experience a force due to this electric field given by Eq. (1.19) and the test electron will move exactly the same as predicted by moving in the magnetic field in the fixed frame of reference.

We therefore see that electric and magnetic fields are not separate and distinct entities, but are really integrated together. A magnetic field in one frame of reference can completely disappear when viewed from another moving frame of reference and be replaced by an electric field that wasn't there in the fixed reference frame.

In 1916, Einstein published his general theory of relativity that explained gravity in terms of bodies warping space. Einstein emmigrated to America in 1933 and would work on his unified field theory at the Institute for Advanced Studies at Princeton until his death in 1955 at the age of 76.

1.5 Electrical Engineering in the Twentieth Century

By the end of the 19th century, **Thomas Edison** had invented the electric light bulb and was lighting cities with his direct current (DC) generating stations. **Nikola Tesla** was a creative inventor (with over 700 patents) and invented the polyphase alternating-current (AC) system, and working with **George Westinghouse**, competed with Edison to generate electricity for widespread use. Tesla designed the first hydroelectric power system at Niagara Falls in 1895 and his AC system of power generation proved to be better than Edison's DC system because AC power could be transmitted over long distances by using transformers to generate high voltages and therefore lower currents. Low currents are important since the line losses are proportional to the square of the current meaning that the lower the current, the lower the line losses for a given power supplied to a load. Among Tesla's many patents were ones related to producing and transmitting radio waves.

The development of the vacuum tube by **Lee DeForest** and others in the early 1900s led to the rapid development of radio as a commercial enterprise. Radios became widely used in World War I and the first radio licenses were issued in 1921. By 1924, two and a half million radio sets were in use in the United States. This number would grow to 33 million by 1936. In 1940, the first TV station went on the air in New York with about 10,000 viewers and NBC aired the first televised newscast in 1944. In 1950, the average home had two radios, and in 1951, one and a half million TV sets were manufactured, a ten-fold increase from the previous year

During the first half of the 20th century while electricity, telephones, and radio were being installed in homes across the country, a quiet revolution was going on in physics. In 1900, **Max Planck**, while trying to explain the frequency spectrum of blackbody radiation, assumed that the energy of light waves could have only certain fixed values that were multiples of $h\nu$, where ν is the frequency and h is a constant equal to 6.63×10^{-34} joule-second, which in now called *Planck's constant*. He needed to make this assumption in order to have his equation agree with the experimental data but he didn't believe that light really behaved this way. In 1905, the same year he published his paper on relativity, Einstein published another paper on the photoelectric effect in which he stated that light does, in fact, come in bundles of energy of size $h\nu$, called a *quantum* of energy. Instead of treating light as a wave, Einstein was treating light as particles, called *photons*. In 1913, **Niels Bohr** developed his model of the hydrogen atom in which a single electron circles a single proton in certain fixed orbits. These orbits are called energy levels and the electron can have only the energy associated with a particular energy level. If the electron gains energy (say be colliding with other particles) it will move to an outer orbit and if it then drops to a lower orbit it will give off energy in the form of a photon with energy $h\nu$. Thus, the frequency of the light emitted by the hydrogen atom depends only on the difference in energy between two energy levels. Bohr's model of the hydrogen atom was able to explain the lines in the hydrogen spectrum that had baffled scientists for years. A similar mechanism is responsible for the different colors of light emitting diodes (LEDs).

For the next 15 years, many scientists including **Wolfgang Pauli**, **Louis de Broglie**, **Werner Heisenberg**, and **Erwin Schrödinger**, building on Bohr's model of atoms, developed what became known as quantum mechanics. This quantum theory, which purports to describe nature at the very small dimensions of atoms and molecules, predicts results every bit as weird as relativity theory. The theory requires you to think of both light and electrons as both waves and particles (depending on what you are thinking about!). Pauli's exclusion principle states that no more than two electrons can occupy the same energy level and that these two electrons must have opposite *spins*. Thus, as electrons are added to atoms to form new elements the electrons must occupy new higher energy levels.

Quantum mechanics led to a better understanding of how electrons in conductors produce currents. When a large number of atoms are brought together to form a crystalline solid, the energy levels of the individual atoms merge to form bands in which the difference between individual energy levels is so small that, for all practical purposes, an electron can have any energy value within the band. A given material will typically have a number of energy bands separated by forbidden bands. The outermost band that is occupied by electrons is called the *valence band*. If this valence band is full and the forbidden gap to the next allowable band is large, then the material will be an insulator such as diamond. On the other hand, if the valence band is not full so that electrons are free to move with very small changes in energy, then the material is a conductor such as copper. If the valence band is full but the forbidden gap to the next available band is small, the material is a semiconductor such as silicon.

In 1947-1948, **William Brattain**, **John Bardeen**, and **William Shockley** invented the transistor. This meant that the large power-hungry vacuum tubes could be replaced with small, lower-powered transistors. In 1958, **Jack Kilby** and independently **Robert Noyce** invented the integrated circuit that allowed many transistors to be produced on a single chip of silicon.

In 1965, **Gordon Moore** predicted that the number of transistors in an integrated circuit would continue to double each year to 65,000 in 1975 as shown in Fig. 1.16. A decade later in 1975, he modified his "law" and predicted that the number of components would double every two years. This Moore's law has proved to be quite accurate over the past 35 years as shown in Fig. 1.17.

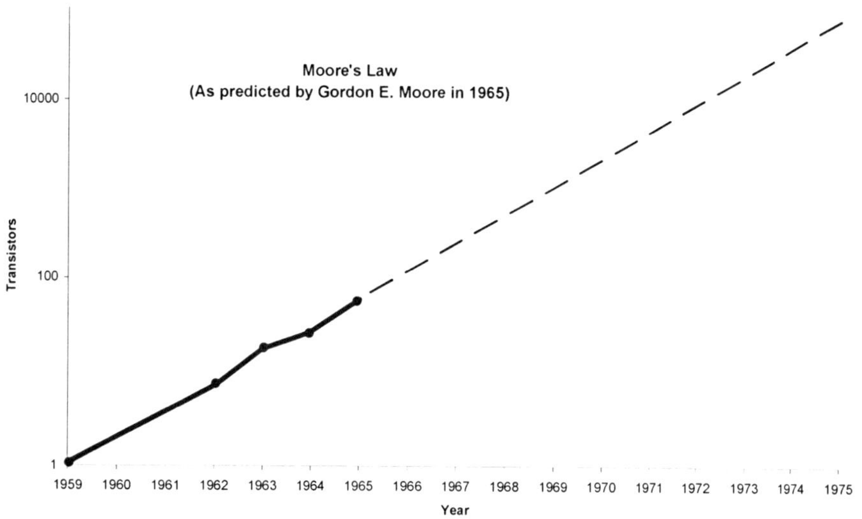

Figure 1.16 Moore's original prediction in 1965

Moore's law as demonstrated in Fig. 1.17 is remarkable. Starting with one transistor on a chip in 1959, engineers were able to put 1 million transistors on a single chip by 1989, 4 million by 1992, 100 million by 2000, and today chips can be made with upwards of 1000 million, or a billion transistors per chip. What to do with all of these transistors? What has happened over the past 25 years is that engineers have put entire computers and other advanced digital circuits on these chips. In many ways, this has changed electrical engineering from primarily an analog circuit's world to one in which almost all products today, including radios, TVs, cell phones, automobiles, and iPods, have substantial digital components. A companion book entitled *Basic Digital Design Using Verilog* introduces you to the basics of designing digital circuits.

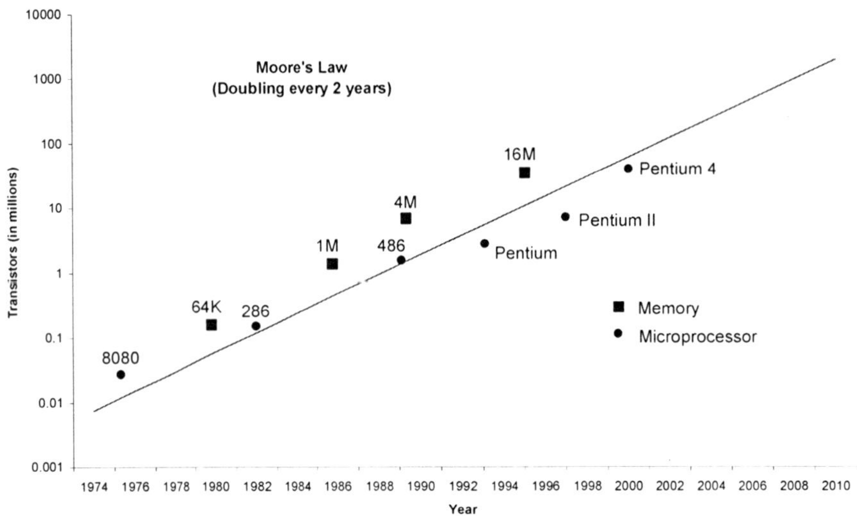

Figure 1.17 Moore's modified prediction in 1975

We have seen that the 18th century was characterized by electrical friction machines and Leyden jars. With the invention of the battery in 1800, the 19th century saw currents producing magnetic fields, changing magnetic fields producing electric fields, and culminating in Maxwell's equations that predicted electromagnetic waves. The first half of the 20th century saw the development of radio and television. The new quantum theory led to the development of semiconductors and the invention of the transistor. The second half of the 20th century was the digital age with the development of computers, microprocessors, and programmable logic devices.

What will the 21st century bring? Based on our historical look at electricity over the past 300 years, you can expect even more changes in the future. While Einstein's theory of general relativity and quantum mechanics were giant developments in the 20th century, the dirty little secret in physics is that they can't both be correct. New physical theories, such as string theory, are actively being pursued in a quest for a theory of everything that will resolve the current contradictions in current physical theories. You should expect such advances to be made in the coming years and, when they do, it is likely that new horizons in electrical engineering will open up that are unimaginable today. Already people are working on quantum computers and nanotechnology. Biological computing is not out of the question in the future. So what should you do to be ready for all of the exciting opportunities that lie ahead? You can begin by learning to calculate the voltages and currents in electrical circuits.

Further Reading

William Gilbert, *De Magnete*, 1600; Translated from Latin by P. Fleury Mottelay, 1893; published by Dover Publications, New York, 1958.

H. W. Brands, *The First American – The Life and Times of Benjamin Franklin*, Doubleday, New York, 2000.

Walter Isaacson, *Benjamin Franklin – An American Life*, Simon & Schuster, New York, 2003.

Richard P. Feynman, Robert B. Leighton, and Matthew Sands, *The Feynman Lectures on Physics, Vol. II – Mainly Electromagnetism and Matter*, Addison-Wesley, Reading, MA, 1964.

Morris H. Shamos, Ed., *Great Experiments in Physics – Firsthand Accounts from Galileo to Einstein*, originally published by Holt, Rinehart and Winston, New York, 1959; republished by Dover Publications, New York, 1987.

Alan Hirshfeld, *The Electric Life of Michael Faraday*, Walker & Co., New York, 2006.

Albert E. Moyer, *Joseph Henry – The Rise of an American Scientist*, Smithsonian Institution Press, Washington, 1997.

Basil Mahon, *The Man Who Changed Everything – The Life of James Clerk Maxwell*, Wiley, Chichester, England, 2003.

James Clerk Maxwell, *An Elementary Treatise on Electricity*, 2nd Ed. originally published by The Clarendon Press, Oxford, 1888; republished by Dover Publications, New York, 2005.

Walter Isaacson, *Einstein – His Life and Universe*, Simon & Schuster, New York, 2007.

George Gamow, *Thirty Years That Shook Physics – The Story of Quantum Theory*, Dover Publications, New York, 1966.

Problems

1.1   The figure below shows two parallel wires carry currents i_1 and i_2 both into the page.

    (a) Show the magnetic fields B_1 and B_2 generated respectively by these currents.

    (b) Show the direction of the force F_1 on the wire carrying i_2 due to the current i_1, and also show the direction of the force F_2 on the wire carrying i_1 due to the current i_2. Determine if the wires will be repelled or attracted.

    (c) If the direction of i_2 is changed so that it flows out of the page, describe what changes.

Chapter 2

Circuit Elements

2.1 Resistors and Ohm's Law

The resistivity ρ is a material property that is a function of temperature and is measured in units of ohm-meters. As shown in Fig. 2.1, the resistance of a circular rod of length l and cross-sectional area A is given by

$$R = \rho l / A \tag{2.1}$$

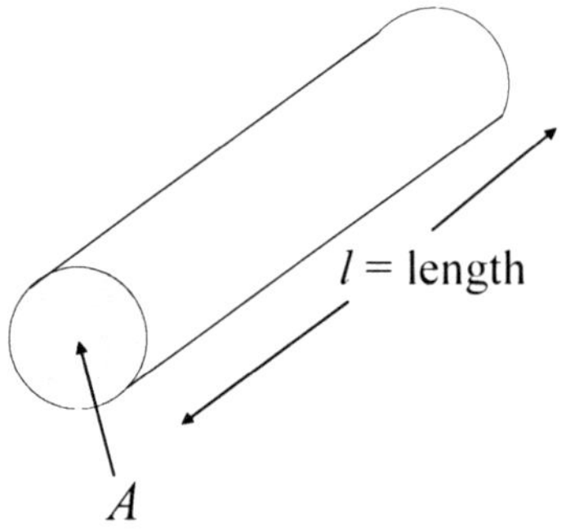

Note that the unit of resistance is the ohm which is written as an upper-case Greek omega, Ω.

The conductivity σ is a material property that is equal to $1/\rho$ and is measured in units of siemens per meter $(\mathrm{S \cdot m^{-1}})$.

Fig. 2.2 shows a current i flowing through a resistor R where the voltage across the resistor is v. We use the *passive sign convention* in which the current i flows from a higher voltage to a lower voltage, i.e. it enters the resistor on the + side of the voltage.

If this is the case, then we can write Ohm's law as (see Section 1.3)

l = length

A

Figure 2.1 Resistance of a circular rod is $R = \rho l / A$

$$v = iR \qquad i = \frac{v}{R} \qquad R = \frac{v}{i} \tag{2.2}$$

The direction that we draw the current, and where we put the + and − signs on the voltage, are completely arbitrary. For example, if we redraw Fig. 2.2a as shown in Fig. 2.2b, and if nothing else has changed, then we must have $i = -i_1$ since otherwise it would not make sense; someone assuming current flow in one direction would get a different result than someone assuming current flow in the opposite direction. Replacing i in Eq. (2.2) with $-i_1$ yields

$$v = -i_1 R \qquad i_1 = -\frac{v}{R} \qquad R = -\frac{v}{i_1} \tag{2.2a}$$

Thus, we say that when using our passive sign convention we would write Ohm's law as $v = -i_1 R$.

(a)   (b)

Figure 2.2 Illustrating Ohm's Law

There are two special cases that are very important. A value of $R = 0$ corresponds to a *short circuit* and a value of $R = \infty$ corresponds to *an open circuit*. For a short circuit, the voltage is always zero while the current is arbitrary, whereas for an open circuit the current is always zero while the voltage is arbitrary.   By arbitrary, we mean that the value depends on what elements comprise the rest of the circuit and on the way the circuit is connected.

A resistor can also be described by its conductance G, which is the reciprocal of resistance. That is,

$$G = \frac{1}{R} \tag{2.3}$$

Using the values shown in Fig. 2.3, Ohm's law can be written in terms of conductance as

$$v = \frac{i}{G} \qquad\qquad i = Gv \qquad\qquad G = \frac{i}{v} \tag{2.4}$$

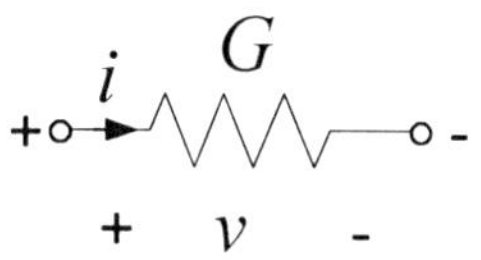

i G

$+$ v $-$

Figure 2.3 Illustrating Ohm's Law
using conductance

Conductance is measured in units of *siemens* (S). Note from Eq. (2.4) that 1 siemen is equal to 1 ampere/volt.

Power (p), in watts (W), is the time rate of expending or absorbing energy (w) in joules. That is,

$$p = \frac{dw}{dt} \tag{2.5}$$

Using the definitions of voltage and current given in Eqs. (1.3) and (1.4), and the passive sign convention, we can write

$$p = \frac{dw}{dt} = \left[\frac{dw}{dq}\right]\left[\frac{dq}{dt}\right] = vi \tag{2.6}$$

If $p > 0$, power is *absorbed* by the element. If $p < 0$, power is *supplied* by the element to the remainder of the circuit.

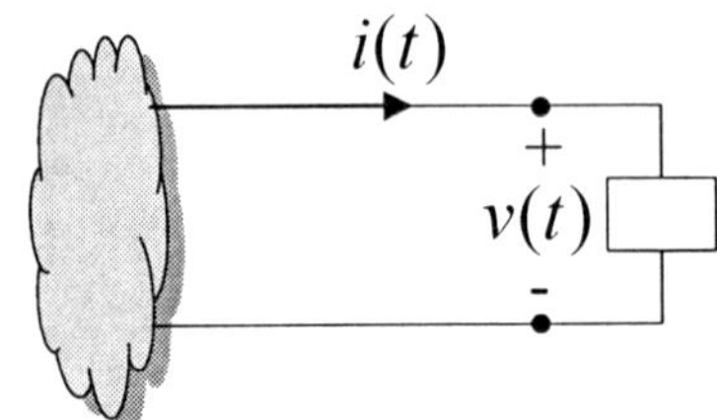

Figure 2.4 The passive sign convention used to
compute the power $p = vi$

We can compute the change in energy from time $t1$ to time $t2$ by integrating Eq. (2.6).

$$w = \int_{t_1}^{t_2} pdt = \int_{t_1}^{t_2} vidt \qquad (2.7)$$

A resistor always dissipates energy. That is, it transforms electrical energy and dissipates it in the form of heat. The rate of energy dissipation is the instantaneous power which, by assuming the passive sign convention and using Ohm's law, can be written as

$$p(t) = v(t)i(t) = Ri^2(t) = \frac{v^2(t)}{R} \geq 0 \qquad (2.8)$$

or, in terms of conductance,

$$p(t) = v(t)i(t) = Gv^2(t) = \frac{i^2(t)}{G} \geq 0 \qquad (2.9)$$

On many common resistors the value of the resistor is color coded using four color bands as shown in Fig. 2.5.

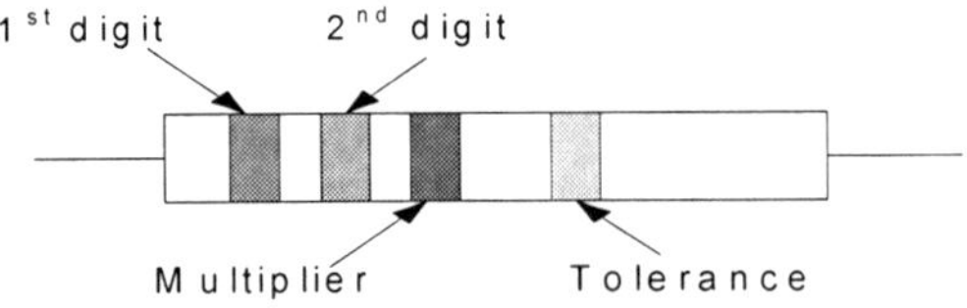

Figure 2.5 Color codes on a resistor

The numbers associated with each color are shown in Table 2.1.

Table 2.1 Resistor Color Codes

Black	Brown	Red	Orange	Yellow	Green	Blue	Violet	Gray	White
0	1	2	3	4	5	6	7	8	9

A gold tolerance band indicates a 5% tolerance.

For example, in Fig. 2.5 if the first digit band is green, you would write a 5; if the second digit band is blue, you would write a 6; if the multiplier band is red, you would add two zeros. Thus, the value of the resistor would be 5600 ohms, or 5.6 kΩ.

In addition to the 4-band resistor code shown in Fig. 2.5 there is a 5-band code that includes a third digit before the multiplier band. This 5-band code is often used for 1% resistors that have a brown tolerance band. There are many interactive aids on the internet to help you interpret resistor color codes, and find standard resistor values. Just search for *resistor codes*.

Example 1 – Ohm's law: voltage sources

In the circuit shown in Fig. 2.6, if $r1 = 2\Omega$ and $r2 = 4\Omega$, find the two currents, $i1$ and $i2$.

Inasmuch as we know the values of $v1$, $v2$, and $v3$, we can simply use Ohm's law to find the two currents, $i1$ and $i2$ as shown in Matlab Example 1.

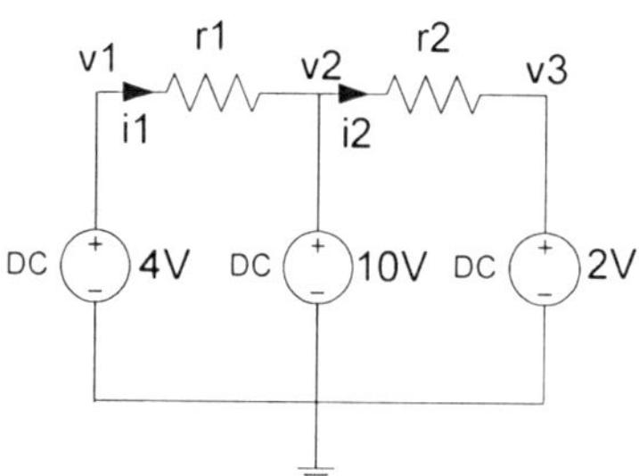

Figure 2.6 Example 1

Matlab Example 1

```
>> v1 = 4;
>> v2 = 10;
>> v3 = 2;
>> r1 = 2;
>> r2 = 4;
>> i1 = (v1-v2)/r1
i1 =

    -3

>> i2 = (v2-v3)/r2
i2 =

     2
```

Note that $i1 = -3$A. This means that the current actually flows in the opposite direction of the $i1$ arrow in the circuit; i.e. the current flows from $v2$ to $v1$. On the other hand, the current $i2 = 2$A. This means that the current flows in the direction of the $i2$ arrow in the circuit; i.e. the current flows from v2 to v3.

Example 2 – Ohm's law: current sources

In the circuit shown in Fig. 2.7, if $r1 = 2\Omega$ and $r2 = 4\Omega$, find the two voltages, $v1$ and $v3$.

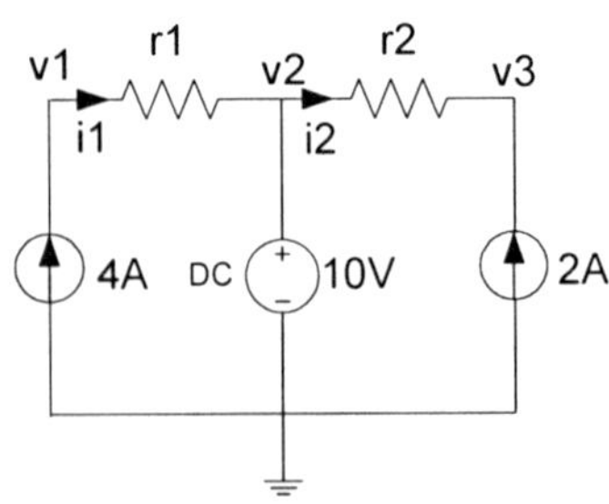

Figure 2.7 Example 2

From Ohm's law, the voltage across resistor $r1$ is $v1 - v2 = i1 \times r1$ and the voltage across resistor $r2$ is $v2 - v3 = i2 \times r2$. From the circuit, $v2 = 10V$, $i1 = 4A$, and $i2 = -2A$ since $i2$ is in the opposite direction to the current through the 2A current source and r2 must carry the same current as the current source. Thus, from the following Matlab solution, both $v1$ and $v3$ are 18V.

Matlab Example 2

```
>> r1 = 2;
>> r2 = 4;
>> v2 = 10;
>> i1 = 4;
>> i2 = -2;
>> v1 = v2 + i1*r1
v1 =
      18

>> v3 = v2 - i2*r2
v3 =
      18
```

2.2 Series and Parallel Resistors

Consider the two resistors R_1 and R_2 in series as shown in Fig. 2.8a. Using Kirchhoff's voltage law and Ohm's law, we can write

$$v_0 = v_1 + v_2 = iR_1 + iR_2$$
$$= i(R_1 + R_2) = iR_S \tag{2.10}$$

We therefore see that we can replace the two resistors R_1 and R_2 in series with a single resistor R_S given by

$$R_S = R_1 + R_2 \tag{2.11}$$

as shown in Fig. 2.8b.

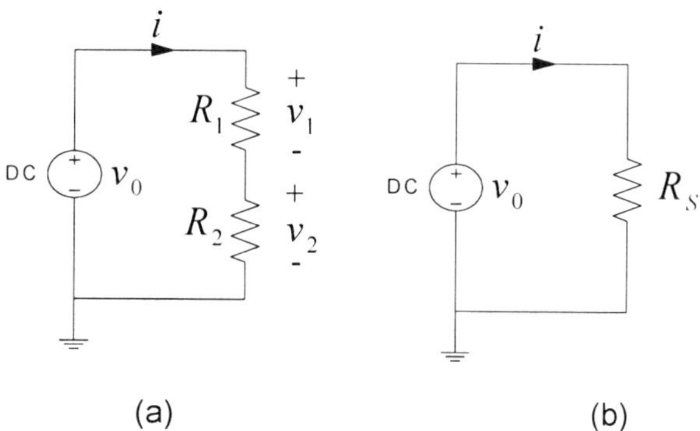

Figure 2.8 Series resistors

The equivalent resistance R_S for N resistors in series will be given by

$$R_S = R_1 + R_2 + \ldots + R_N \tag{2.12}$$

Now consider the two resistors R_1 and R_2 in parallel as shown in Fig. 2.9a. Using Kirchhoff's current law and Ohm's law we can write

$$i = i_1 + i_2 = \frac{v}{R_1} + \frac{v}{R_2}$$

$$= v\left(\frac{1}{R_1} + \frac{1}{R_2}\right) = \frac{v}{R_P} \tag{2.13}$$

We therefore see that we can replace the two resistors R_1 and R_2 in parallel with a single resistor R_P given by

$$\frac{1}{R_P} = \frac{1}{R_1} + \frac{1}{R_2} \tag{2.14}$$

which can be rewritten as

$$R_P = \frac{R_1 R_2}{R_1 + R_2} \tag{2.15}$$

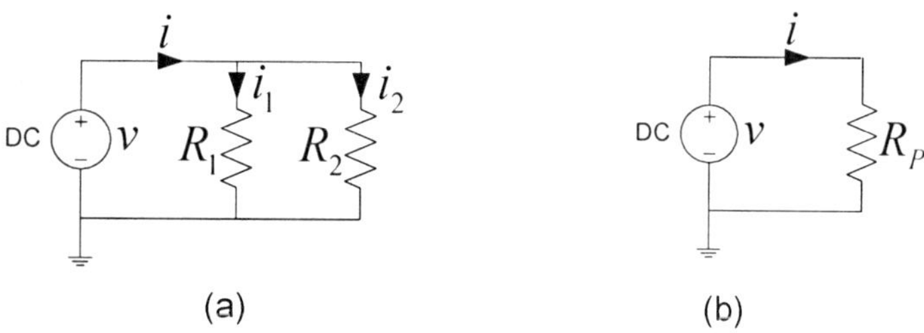

Figure 2.9 Parallel resistors

The equivalent resistance R_P for N resistors in parallel will be given by

$$\frac{1}{R_P} = \frac{1}{R_1} + \frac{1}{R_2} + \ldots + \frac{1}{R_N} \tag{2.16}$$

Example 3 – Equivalent Resistances

Consider the circuit shown in Fig. 2.10. If $r1 = r2 = r3 = r4 = r5 = 2\Omega,$ find the equivalent resistance r.

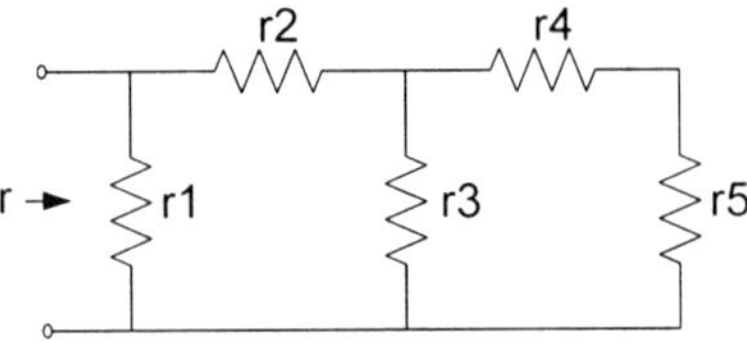

Figure 2.10 Example 3

Note that $r4$ and $r5$ are in series and thus can be replaced with an equivalent resistance $r6 = r4 + r5$ as shown in Fig 2.11.

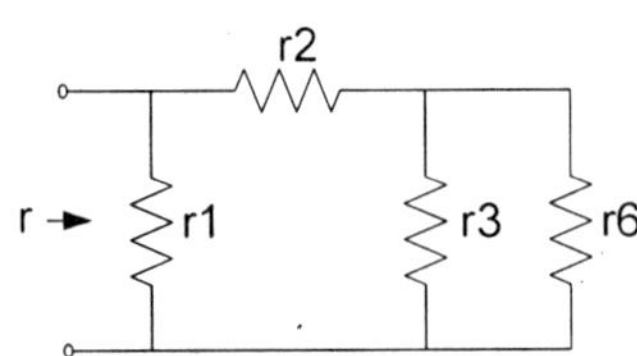

Figure 2.11 Example 3 – step 1

Now $r3$ and $r6$ are in parallel and can be replaced with an equivalent resistance $r7 = \dfrac{r3 \times r6}{r3 + r6}$ as shown in Fig 2.12.

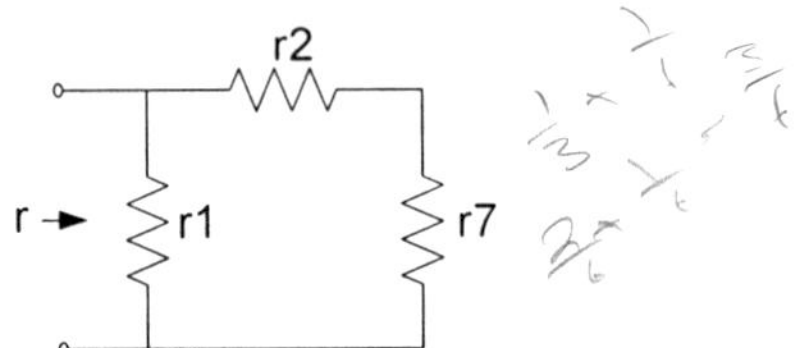

Figure 2.12 Example 3 – step 2

Now $r2$ and $r7$ are in series and can be replaced with an equivalent resistance $r8 = r2 + r7$ as shown in Fig 2.13.

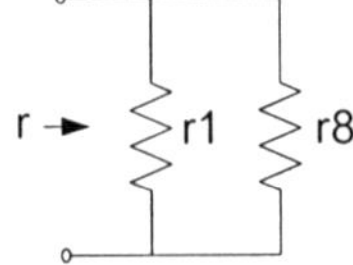

Figure 2.13 Example 3 – step 3

Now $r1$ and $r8$ are in parallel and can be replaced with the final equivalent resistance $r = \dfrac{r1 \times r8}{r1 + r8} = 1.25\ \Omega$ as shown in Matlab Example 3.

Matlab Example 3

```
>> r1=2;
>> r2=2;
>> r3=2;
>> r4=2;
>> r5=2;
>> r6 = r4 + r5
r6 =
        4

>> r7 = (r3*r6)/(r3+r6)
r7 =
     1.3333

>> r8 = r2 + r7
r8 =
     3.3333

>> r = (r1*r8)/(r1+r8)
r =
     1.2500
```

2.3 Capacitors

We saw in Chapter 1 that a capacitor is a passive element that stores energy in its electric field. A capacitor can be represented by two conducting plates separated by an insulator (or dielectric) as shown in Fig. 2.14. When a voltage source is connected to the capacitor, the source deposits a positive charge, $+q$, on one plate and a negative charge, $-q$, on the other. The amount of charge is directly proportional to the voltage so that

$$q = Cv \qquad (2.17)$$

where C is the *capacitance* that depends only on the geometry of the capacitor and the properties of the dielectric material between the plates. For the parallel plate capacitor shown in Fig. 2.14, the capacitance is given by

$$C = \frac{\varepsilon A}{d} \qquad (2.18)$$

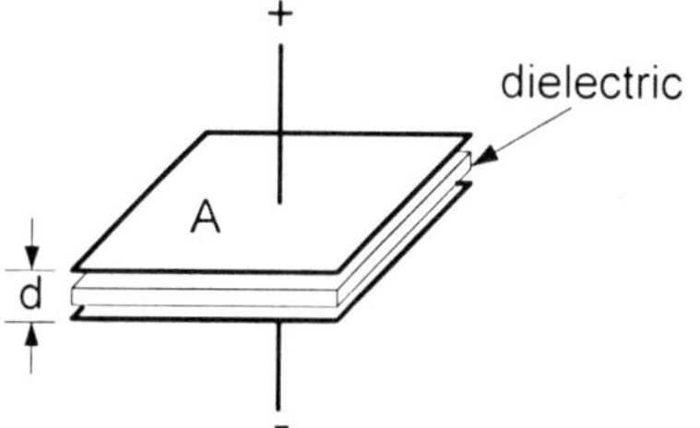

Figure 2.14 Parallel plate capacitor

where A is the area of the plates, d is the distance between the plates, and ε is the permittivity of the dielectric material between the conducting plates. Note that the capacitance increase as the area of the plates increases and the distance between the plates decreases.

The symbols used for a capacitor are shown in Fig. 2.15. Capacitance is measured in *farads* (F) where one farad is equal to 1 coulomb per volt. Typical capacitors have values of capacitance in the pF or µF range.

From Eq. (1.16) in Chapter 1, recall that the current through a capacitor is given by

$$i(t) = C \frac{dv(t)}{dt} \qquad (2.19)$$

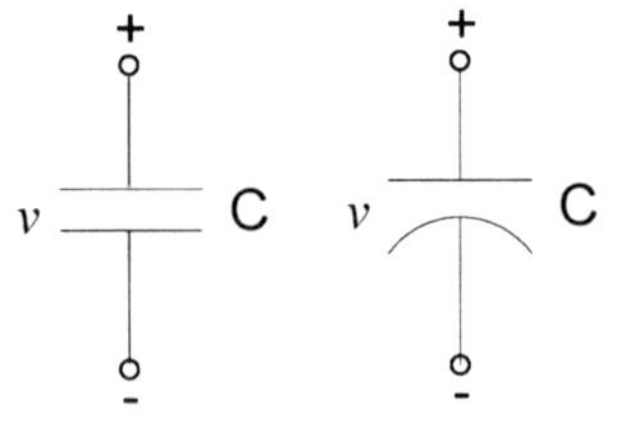

Integrating Eq. (2.19) we can write

Figure 2.15 Symbols for a capacitor

$$v(t) = \frac{1}{C} \int_{-\infty}^{t} i(t)dt = \frac{1}{C} \int_{-\infty}^{t_0} i(t)dt + \frac{1}{C} \int_{t_0}^{t} i(t)dt \qquad (2.20)$$

Note that

$$\frac{1}{C} \int_{-\infty}^{t_0} i(t)dt = v(t_0) = \frac{q(t_0)}{C} \qquad (2.21)$$

where $v(t_0)$ is the voltage at time t_0 and we have assumed that the capacitor is uncharged at time $t = -\infty$ so that $v(-\infty) = 0$. This assumption makes sense since all real capacitors have a leakage resistance, so that any capacitor that is charged and then disconnected from the source and left to sit for a long time will discharge completely. Therefore, from Eqs. (2.20) and (2.21), we can write

$$v(t) = v(t_0) + \frac{1}{C} \int_{t_0}^{t} i(t)dt \qquad (2.22)$$

From the above, we see that the voltage across a capacitor can not change instantaneously since for $v(t_0^+)$ where $t_0^+ = t_0 + \tau$, τ very small, to be different than $v(t_0)$ the current would have to be infinite, which is physically impossible.

The instantaneous power delivered to the capacitor is

$$p(t) = vi = Cv \frac{dv}{dt} \qquad (2.23)$$

We can find the energy stored in the capacitor by integrating Eq. (2.23). Thus,

$$w = \int p(t)dt = C \int_{-\infty}^{t} v \frac{dv}{dt} dt = C \int_{-\infty}^{t} v \, dv$$

from which

$$w = \frac{1}{2} Cv^2(t) \text{ joules} \qquad (2.24)$$

Eq. (2.24) represents the energy stored in the electric field established between the two plates of the capacitor. This energy can be retrieved. In fact, the word capacitor is derived from this element's ability (or capacity) to store energy.

A capacitor has the following important properties:

1. When the voltage across a capacitor is constant (not changing with time) the current through the capacitor $i = C\ dv/dt = 0$. Thus, *a capacitor is an open circuit to dc.* If, however, a dc voltage is suddenly connected across a capacitor, the capacitor begins to charge and store energy.

2. *The voltage across a capacitor cannot change instantaneously.*

3. *An ideal capacitor does not dissipate energy.* It takes power from the circuit when the capacitor is charging (the magnitude of the voltage is increasing) and delivers power to the circuit when the capacitor is discharging (the magnitude of the voltage is decreasing).

4. A real capacitor has a "leakage resistance" which can be modeled as shown in Fig. 2.16. The typical leakage resistance might be 100 MΩ. We will ignore the leakage resistance of capacitors in this book.

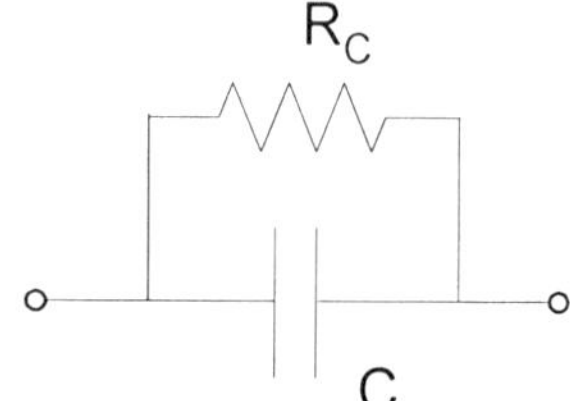

Figure 2.16 Model of a real capacitor

We see from Eq. (2.19) that there is only a current through a capacitor when the voltage across the capacitor is changing. But what does it mean to have a current flowing through the capacitor? We are assuming that we have a perfect dielectric so no charges actually flow between the plates of a capacitor. When the voltage on the upper plate in Fig. 2.14 is increasing, this plate becomes more positively charged (as electrons are actually flowing away from it into the circuit) and, when the voltage on the upper plate is decreasing, the plate becomes less positively charged (as electrons are actually flowing in from the rest of the circuit). The charge on the lower plate is always the negative of the charge on the upper plate.

But what about the space between the parallel plates of the capacitor? Does a current flow there? Remember from Chapter 1 that there will be an electric field between these parallel plates and, when the voltage across the capacitor changes, this electric field will change. But in Section 1.4 of Chapter 1, we saw that Maxwell introduced the new *displacement current* term $\partial \mathbf{D}/\partial t$ into his electrodynamics equations. (This was the term that gave rise to electromagnetic waves.) Thus, a changing electric field produces a displacement current, even in free space, that behaves in every way like a current produced by moving charges. Therefore, in a circuit containing a capacitor the current through the wires of the circuit continues in the dielectric of a capacitor as a displacement current!

Example 4 – Sinusoidal Voltage on a Capacitor

Suppose that the voltage $v_0 = 5\sin\omega t$ is applied across a capacitor C as shown in Fig. 2.17. If the frequency of the sine wave is 10 kHz and the capacitance is 10 µF plot the current through the capacitor and the power delivered to the capacitor as a function of time for two cycles of the voltage sine wave.

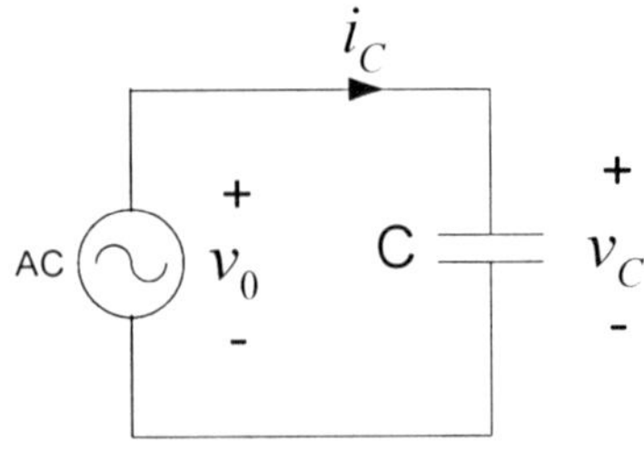

Figure 2.17 Example 4

The angular frequency ω in radians is given by $\omega = 2\pi f = 2\pi/T$ where f is the frequency in Hz and T is the period in seconds. Note that $v_C = v_0 = 5\sin(2\pi t/T)$ and

$$i_C(t) = C\frac{dv(t)}{dt} = \frac{2\pi C 5}{T}\cos\left(\frac{2\pi t}{T}\right) \tag{2.25}$$

From Eq. (2.23), the instantaneous power delivered to the capacitor is

$$p(t) = v_C(t)i_C(t) = \frac{50\pi C}{T}\sin\left(\frac{2\pi t}{T}\right)\cos\left(\frac{2\pi t}{T}\right) \tag{2.26}$$

The Matlab solution is shown in Matlab Example 4 and the resulting plot is shown in Fig. 2.18.

Matlab Example 4

```
>> C = 10*10^-6;
>> f = 10000;
>> T = 1/f;
>> t = linspace(0,2*T,100);
>> vc = 5*sin(2*pi*t/T);
>> ic = ((10*pi*C)/T)*cos(2*pi*t/T);
>> p = vc.*ic;
>> curves = [vc; ic; p];
>> plot(t,curves)
```

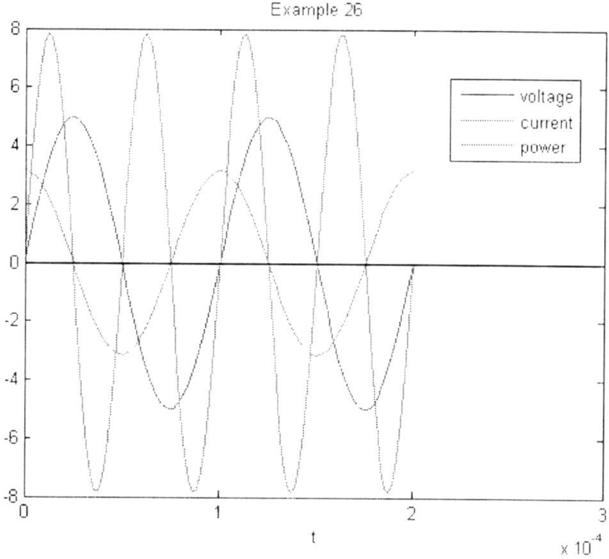

Figure 2.18 Plot resulting from Matlab Example 4

So far, we have used Matlab as a simple calculator. However, the variables in Matlab are generally vectors or matrices (see Appendix D). In Matlab Example 4, the statement

$$t = \texttt{linspace(0,2*T,100);}$$ (2.27)

will generate 100 equispaced elements in the row vector t starting at 0 and ending with 2*T. This will generate 100 samples in a time interval equal to two periods.

To compute a value of vc for each value of t using the equation $v_C = 5\sin(2\pi t/T)$, we write the Matlab statement

$$\texttt{vc = 5*sin(2*pi*t/T);}$$ (2.28)

If we multiply the row matrix t by 2*pi, each element of t will be multiplied by 2*pi, which is what we want. Similarly, dividing that resulting row matrix by T will divide each element by T, producing a row matrix for the argument of the function *sin*. This *sin* function is itself a row matrix with each element being the *sine* of the corresponding argument of t. Multiplying this *sin* function by 5 will multiply each element of the row matrix by 5, giving the row matrix vc.

In a similar way, the Matlab statement

$$\texttt{ic = ((10*pi*C)/T)*cos(2*pi*t/T);}$$ (2.29)

will compute the current $i_C(t)$ given by Eq. (2.25). However, when we want to compute the power

$$p(t) = v_C(t)i_C(t) \qquad\qquad (2.30)$$

we can't just multiply Eqs. (2.28) and (2.29) because these are row matrices and Matlab will try to do a matrix multiplication which is not possible with two row maticies. What we want to do is to multiply each element of *vc* in Eq. (2.28) by the corresponding element of *ic* in Eq. (2.29). The *dot-multiply operator* .* will do this. Dot operators perform element-by-element operations. In addition to the .* operator, the operators ./ and .^ will perform element-by-element division and exponentiation respectively. Therefore, the Matlab equation

```
p = vc.*ic;
```
$$\qquad (2.31)$$

will produce a row matrix containing the power as a function of time.

To plot three curves (voltage, current, and power) on the same graph, we need to store the voltage, current, and power vectors in a single 3 x 40 matrix. The statement

```
curves = [vc; ic; p];
```
$$\qquad (2.32)$$

will do this. Then the statement

```
plot(t,curves)
```
$$\qquad (2.33)$$

will plot the three curves shown in Fig. 2.18.

Note in Fig. 2.18 that the current and voltage waveforms are out of phase by 90 degrees. That is, the current waveform reaches its maximum value 90 degrees before the voltage waveform reaches its maximum value. We say that the current *leads* the voltage by 90 degrees. (Or the voltage lags the current by 90 degrees.) Also note that when the power is positive, energy is being stored in the capacitor, and when the power is negative, the energy is being returned to the circuit. This power cycle takes place twice for each cycle of the voltage and current, which is equivalent to saying that the instantaneous power varies at twice the frequency of the voltage and current.

2.4 Capacitors in Series and Parallel

Consider N capacitors in parallel as shown in Fig. 2.19a. The currents through the N capacitors are given by

$$i_1 = C_1\frac{dv}{dt} \qquad i_2 = C_2\frac{dv}{dt} \qquad \ldots \qquad i_N = C_N\frac{dv}{dt} \qquad (2.34)$$

Applying KCL (see Section 1.3), we can write

$$i = i_1 + i_2 + \cdots + i_N = \left(C_1 + C_2 + \cdots + C_N \right)\frac{dv}{dt} = C_{eq}\frac{dv}{dt} \qquad (2.35)$$

where C_{eq} is the equivalent capacitance shown in Fig. 2.19b and given by

$$C_{eq} = \sum_{k=1}^{N} C_k \qquad (2.36)$$

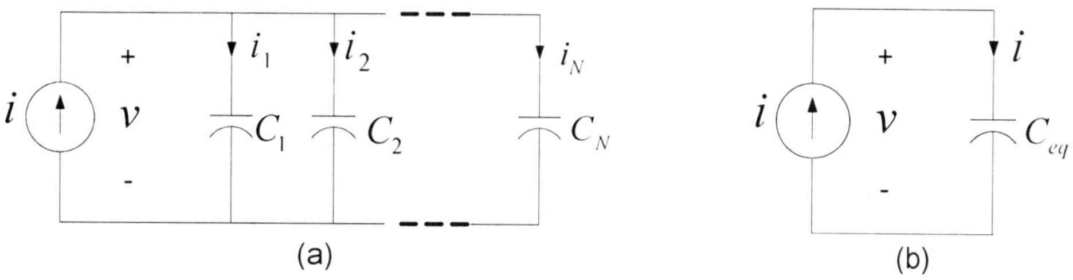

Figure 2.19 Parallel capacitors

Thus, the equivalent capacitance of N capacitors in parallel is the sum of the individual capacitances. We see that *capacitors in parallel act like resistors in series.* Now consider N capacitors in series as shown in Fig. 2.20a.

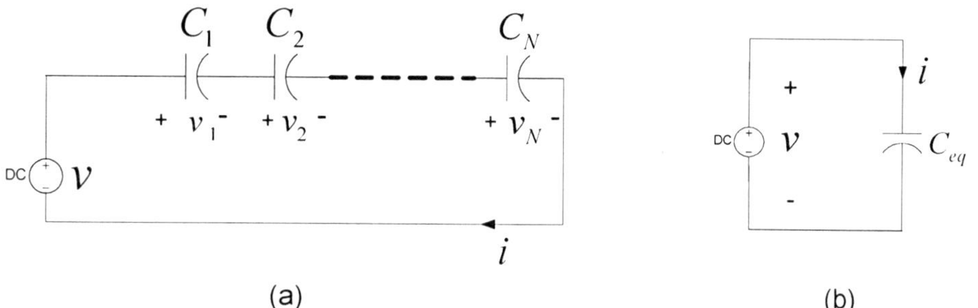

Figure 2.20 Series capacitors

The voltages across the N capacitors are given by

$$v_1 = \frac{1}{C_1}\int i\,dt \quad v_2 = \frac{1}{C_2}\int i\,dt \quad \cdots \quad v_N = \frac{1}{C_N}\int i\,dt \qquad (2.37)$$

Applying KVL (see Section 1.3), we can write

$$v = v_1 + v_2 + \cdots + v_N = \left(\frac{1}{C_1} + \frac{1}{C_2} + \cdots + \frac{1}{C_N} \right)\int i\,dt = \frac{1}{C_{eq}}\int i\,dt \qquad (2.38)$$

where C_{eq} is the equivalent capacitance shown in Fig. 2.20b and given by

$$\frac{1}{C_{eq}} = \sum_{k=1}^{N} \frac{1}{C_k} \qquad (2.39)$$

Thus, the equivalent capacitance of N series connected capacitors is the reciprocal of the sum of the reciprocals of the individual capacitors, and we see that *capacitors in series act like resistors in parallel*.

Example 5 – Capacitors in Series and Parallel

Find the equivalent capacitance C_{eq} between terminals a and b in Fig. 2.21 if all six capacitors have a value of 1 μF.

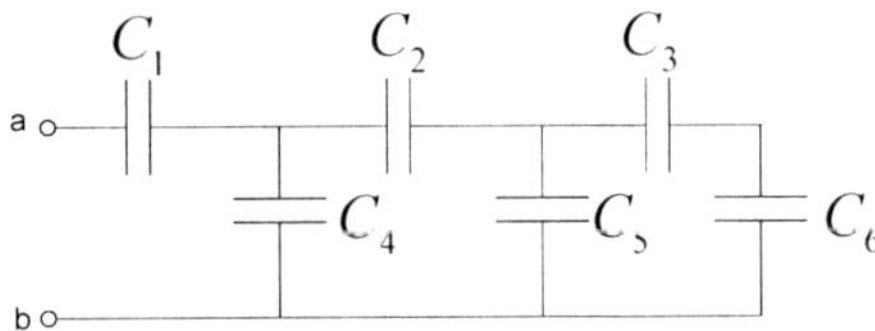

Figure 2.21 Example 27

Capacitors C_3 and C_6 are in series and can therefore be replaced with a single capacitor C_7 as shown in Fig. 2.22a with a value of

$$C_7 = \frac{C_3 C_6}{C_3 + C_6} \qquad (2.40)$$

Now capacitors C_5 and C_7 are in parallel and can therefore be replaced with a single capacitor C_8 as shown in Fig. 2.22b with a value of

$$C_8 = C_5 + C_7 \qquad (2.41)$$

Now capacitors C_2 and C_8 are in series and can therefore be replaced with a single capacitor C_9 as shown in Fig. 2.22c with a value of

$$C_9 = \frac{C_2 C_8}{C_2 + C_8} \qquad (2.42)$$

Now capacitors C_4 and C_9 are in parallel and can therefore be replaced with a single capacitor C_{10} as shown in Fig. 2.22d with a value of

$$C_{10} = C_4 + C_9 \qquad (2.43)$$

Finally, the equivalent capacitance C_{eq} between terminals a and b is just capacitors C_1 and C_{10} in series and therefore has a value of

$$C_{eq} = \frac{C_1 C_{10}}{C_1 + C_{10}} = 0.6154 \ \mu F \qquad (2.44)$$

as shown in Matlab Example 5.

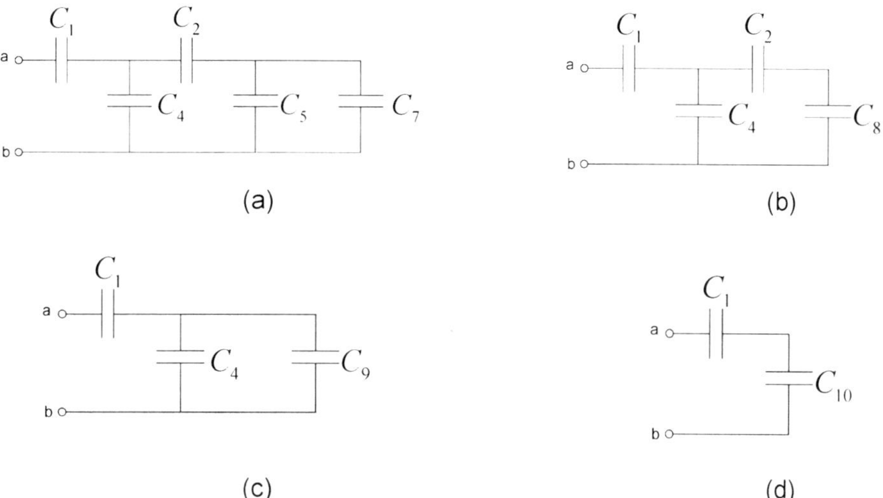

Figure 2.22 Steps to solve Example 5

Matlab Example 5

```
>> c1 = 1;
>> c2 = 1;
>> c3 = 1;
>> c4 = 1;
>> c5 = 1;
>> c6 = 1;
>> c7 = c3*c6/(c3+c6)
c7 =
     0.5000

>> c8 = c5 + c7
c8 =
     1.5000

>> c9 = c2*c8/(c2+c8)
c9 =
     0.6000
>> c10 = c4 + c9
c10 =
     1.6000

>> ceq = c1*c10/(c1+c10)
ceq =
     0.6154
```

2.5 Inductors

We saw in Chapter 1 that an inductor is a passive element that stores energy in its magnetic field. An inductor can be represented by a coil of wire wrapped around a core as shown in Fig. 2.23. The current through the coil will produce a magnetic flux density **B** that is proportional to the current. If the current is changing with time, the magnetic flux density will be changing with time and by Faraday's law this will induce a voltage that will appear across the coil from the + to the − sign in Fig. 2.23. This voltage will be proportional to the time rate of change of the current and given by

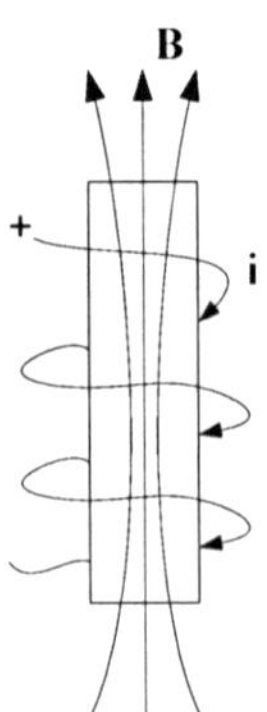

Figure 2.23 Forming an inductor

$$v(t) = L\frac{di(t)}{dt} \tag{2.45}$$

where L is the *inductance* in henrys (H), and 1 H = 1 volt second/ampere. The sign of the induced voltage follows Lenz's law and is always such that the current flowing through an external resistor would be opposite to the current that produced the flux. This is consistent with the passive sign convention shown on the symbol for an inductor in Fig. 2.24.

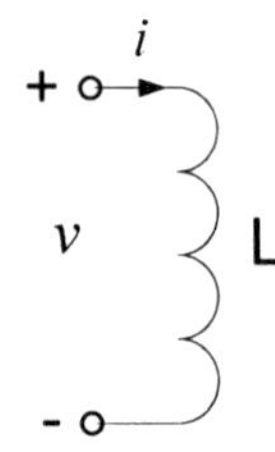

Figure 2.24 Symbol for an inductor

Integrating Eq. (2.45), we can write

$$i(t) = \frac{1}{L}\int_{-\infty}^{t} v(t)dt = \frac{1}{L}\int_{-\infty}^{t_0} v(t)dt + \frac{1}{L}\int_{t_0}^{t} v(t)dt \tag{2.46}$$

Note that

$$\frac{1}{L}\int_{-\infty}^{t_0} v(t)dt = i(t_0) \tag{2.47}$$

where $i(t_0)$ is the current at time t_0 and we have assumed that the current at time $t = -\infty$ is zero. Therefore, from Eqs. (2.39) and (2.40), we can write

$$i(t) = i(t_0) + \frac{1}{L}\int_{t_0}^{t} v(t)dt \tag{2.48}$$

From the above, we see that the current through an inductor can not change instantaneously since, for $i(t_0^+)$ where $t_0^+ = t_0 + \tau$, (τ very small) to be different than $i(t_0)$, the voltage would have to be infinite, which is physically impossible.

The instantaneous power delivered to the inductor is

$$p(t) = vi = Li\frac{di}{dt} \tag{2.49}$$

We can find the energy stored in the magnetic field by integrating Eq. (2.49). Thus,

$$w_L(t) = \int p(t)\,dt = L \int_{-\infty}^{t} i\,\frac{di}{dt}\,dt = L \int_{-\infty}^{t} i\,di$$

from which

$$w_L(t) = \frac{1}{2} L i^2(t) \ \text{joules} \qquad\qquad (2.50)$$

An inductor has the following important properties:

1. When the current through an inductor is constant (not changing with time), the voltage across the inductor $v = L\,di/dt = 0$. Thus, *an inductor acts like a short circuit to dc.*

2. *The current through an inductor cannot change instantaneously.*

3. *An ideal inductor does not dissipate energy.* It takes power from the circuit when the magnetic field is building up (the magnitude of the current is increasing) and delivers power to the circuit when the magnetic field is collapsing (the magnitude of the current is decreasing).

4. A real inductor has a small but finite resistance along the coil and a small "winding capacitance" between coil segments. This can be modeled as shown in Fig. 2.25. We will ignore the coil resistance and capacitance in this book and treat all inductors as ideal inductors.

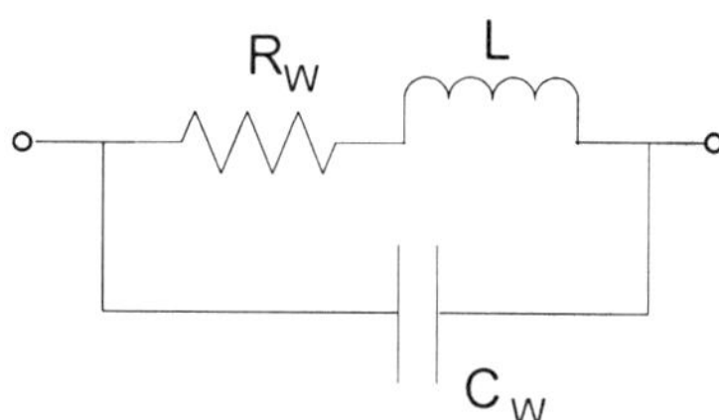

Figure 2.25 Model of a real inductor

Example 6 – Inductors

Suppose that the voltage $v_0 = 5\sin\omega t$ is applied across an inductor L as shown in Fig. 2.26. If the frequency of the sine wave is 1 kHz and the inductance is 0.5 mH, plot the voltage across and the current through the inductor and the power delivered to the inductor as a function of time for two cycles of the voltage sine wave.

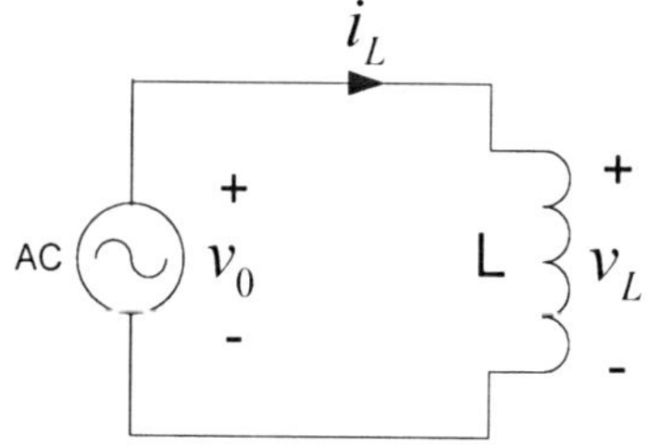

Figure 2.26 Example 6

The angular frequency ω in radians is given by $\omega = 2\pi f = 2\pi/T$ where f is the frequency in Hz and T is the period in seconds. Note that $v_L = v_0 = 5\sin(2\pi t/T)$ and, from Eq. (2.41)

$$i_L(t) = i(t_0) + \frac{1}{L}\int_{t_0}^{t} v_L(t)\,dt$$

or

$$i_L(t) = i(t_0) + \frac{1}{L}\int_{t_0}^{t} 5\sin\left(\frac{2\pi t}{T}\right)dt \qquad (2.51)$$

The lower limit in the integral in Eq. (2.44) is just $i(t_0)$ so that

$$i_L(t) = -\frac{5T}{2\pi L}\cos\left(\frac{2\pi t}{T}\right) \qquad (2.52)$$

From Eq. (2.42), the instantaneous power delivered to the inductor is

$$p(t) = v_L(t)i_L(t) = -\frac{25T}{2\pi L}\sin\left(\frac{2\pi t}{T}\right)\cos\left(\frac{2\pi t}{T}\right) \qquad (2.53)$$

The Matlab solution is shown in Matlab Example 6 and the resulting plot is shown in Fig. 2.27.

Matlab Example 6

```
>> L = 0.5*10^-3;
>> f = 1000;
>> T = 1/f;
>> t = linspace(0,2*T,80);
>> vl = 5*sin(2*pi*t/T);
>> il = -((5*T)/(2*pi*L))*cos(2*pi*t/T);
>> p = vl.*il;
>> curves = [vl; il; p];
>> plot(t,curves)
```

Compare the results in Fig. 2.27 with those in Fig. 2.18 in Example 4. In both cases, the current and voltage waveforms are out of phase by 90 degrees. But in Fig. 2.27, the voltage waveform reaches its maximum value 90 degrees before the current waveform reaches its maximum value. We say that the voltage *leads* the current by 90 degrees. (Or the current *lags* the voltage by 90 degrees.) This is just the opposite of the capacitor in Example 4. Also note that when the power is positive, energy is being stored in the magnetic field (current is either positive and increasing, or negative and decreasing), and when the power is negative, the energy is being returned to the circuit. This power cycle takes place twice for each cycle of the voltage and current.

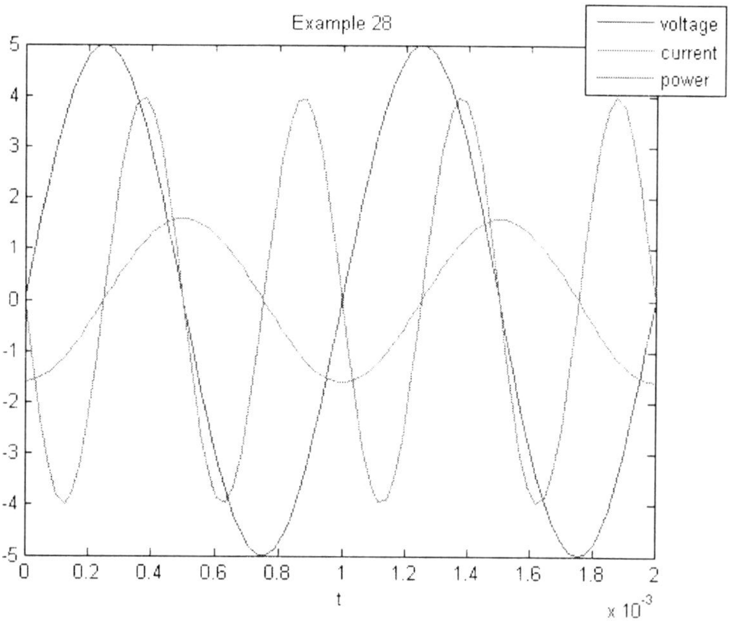

Figure 2.27 Plot resulting from Matlab Example 6

2.6 Inductors in Series and Parallel

Consider N inductors in series as shown in Fig. 2.28a. The voltages across the N inductors are given by

$$v_1 = L_1 \frac{di}{dt} \qquad v_2 = L_2 \frac{di}{dt} \qquad \cdots \qquad v_N = L_N \frac{di}{dt} \qquad (2.54)$$

Applying KVL (see Section 1.3), we can write

$$v = v_1 + v_2 + \cdots + v_N = \left(L_1 + L_2 + \cdots + L_N \right) \frac{di}{dt} = L_{eq} \frac{di}{dt} \qquad (2.55)$$

where L_{eq} is the equivalent inductance shown in Fig. 2.28b and given by

$$L_{eq} = \sum_{k=1}^{N} L_k \qquad (2.56)$$

Thus, the equivalent inductance of series connected inductors is the sum of the individual inductances. We see that *inductances in series combine in the same way as resistors in series*.

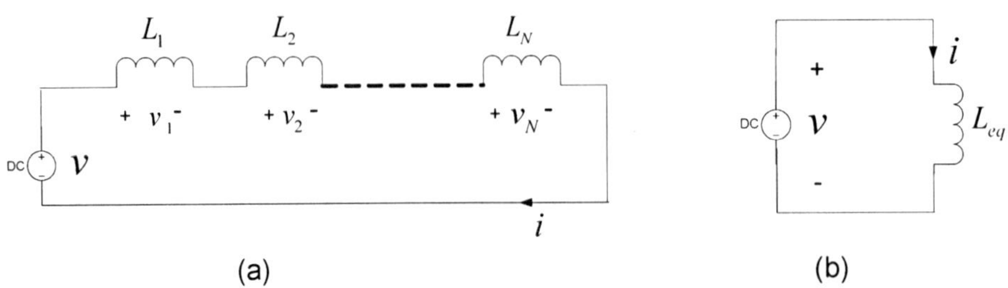

Figure 2.28 Series inductors

Now consider N inductors in parallel as shown in Fig. 2.29a. The currents through the N inductors are given by

$$i_1 = \frac{1}{L_1} \int v\,dt \qquad i_2 = \frac{1}{L_2} \int v\,dt \qquad \dots \qquad i_N = \frac{1}{L_N} \int v\,dt \qquad (2.57)$$

Applying KCL (see Section 1.3) we can write

$$i = i_1 + i_2 + \cdots + i_N = \left(\frac{1}{L_1} + \frac{1}{L_2} + \cdots + \frac{1}{L_N} \right) \int v\,dt = \frac{1}{L_{eq}} \int v\,dt \qquad (2.58)$$

where L_{eq} is the equivalent inductance shown in Fig. 2.29b and given by

$$\frac{1}{L_{eq}} = \sum_{k=1}^{N} \frac{1}{L_k} \qquad (2.59)$$

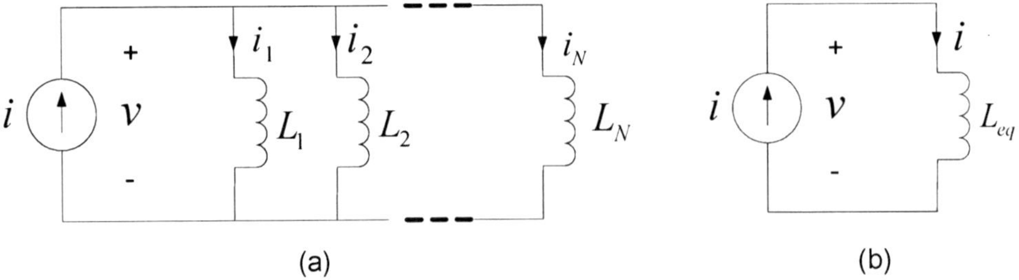

Figure 2.29 Parallel inductors

Thus, the equivalent inductance of N parallel connected inductors is the reciprocal of the sum of the reciprocals of the individual inductances, and we see that *inductances in parallel combine in the same way as resistors in parallel.*

Example 7 – Inductors in Series and Parallel

Find the equivalent inductance L_{eq} between terminals a and b in Fig. 2.30 if all five inductors have a value of 1 mH.

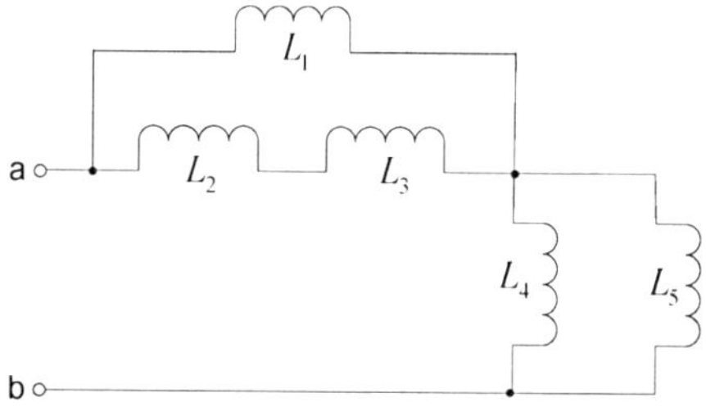

Figure 2.30 Example 7

Inductors L_2 and L_3 are in series and inductors L_4 and L_5 are in parallel. Therefore, inductors L_2 and L_3 can be replaced with a single inductor L_6 with a value of

$$L_6 = L_2 + L_3 \tag{2.60}$$

and inductors L_4 and L_5 can be replaced with a single inductor L_7 with a value of

$$L_7 = \frac{L_4 L_5}{L_4 + L_5} \tag{2.61}$$

as shown in Fig. 7.18a. Now inductors L_1 and L_6 are in parallel and can therefore be replaced with a single inductor L_8 as shown in Fig. 7.18b with a value of

$$L_8 = \frac{L_1 L_6}{L_1 + L_6} \tag{2.62}$$

Finally, the equivalent inductance L_{eq} between terminals a and b is just inductors L_8 and L_7 in series and therefore has a value of

$$L_{eq} = L_8 + L_7 = 1.1667 \text{ mH} \tag{2.63}$$

as shown in Matlab Example 7.

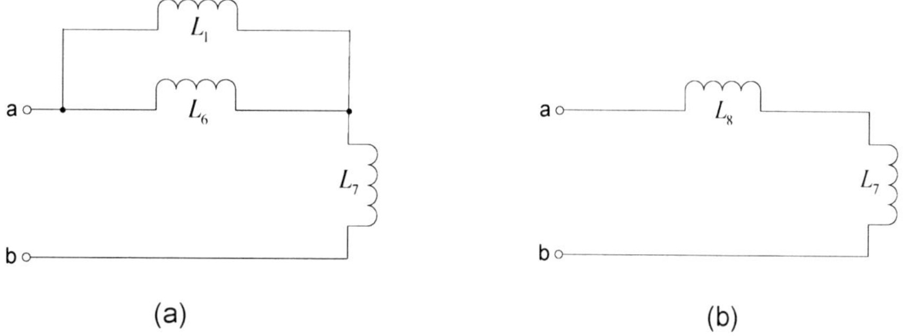

(a)   (b)

Figure 2.31 Steps to solve Example 7

Matlab Example 7

```
>> L1 = 1;
>> L2 = 1;
>> L3 = 1;
>> L4 = 1;
>> L5 = 1;
>> L6 = L2 + L3
L6 =

     2

>> L7 = L4*L5/(L4+L5)
L7 =
     0.5000

>> L8 = L1*L6/(L1+L6)
L8 =
     0.6667

>> Leq = L8 + L7
Leq =
     1.1667
```

Problems

2.1. What is the resistor color code for each of the following resistor values?
  a. $15\ \Omega$
  b. $200\ \Omega$
  c. $3.9\ K\Omega$
  d. $47\ K\Omega$
  e. $560\ K\Omega$
  f. $6.8\ M\Omega$
  g. $82\ M\Omega$

2.2. What is the resistor value for each of the following color codes?
  a. gray-red-black
  b. blue-gray-brown
  c. green-blue-red
  d. yellow-violet-orange
  e. orange-white-yellow
  f. red-black-green
  g. brown-green-blue

2.3 In the circuit to the right, if $r1 = 4\Omega$ and $r2 = 8\Omega$, find the two currents, $i1$ and $i2$.

2.4   In the circuit in Problem 2.3, if $i1 = -1$A and $i2 = -4$A, find the values of the two resistors, $r1$ and $r2$.

2.5   In the figure to the right, if the current i is measured to be 10A and the voltage v is 5V, calculate the power, and determine if the element is absorbing power or supplying power to the rest of the circuit.

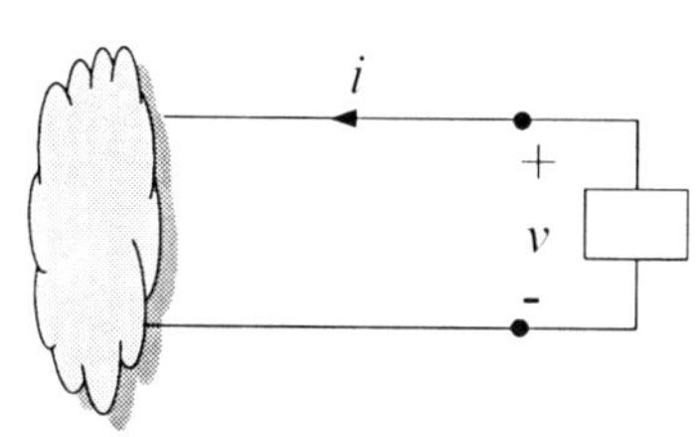

2.6.   In the circuit to the right, find the two currents, $i1$ and $i2$.

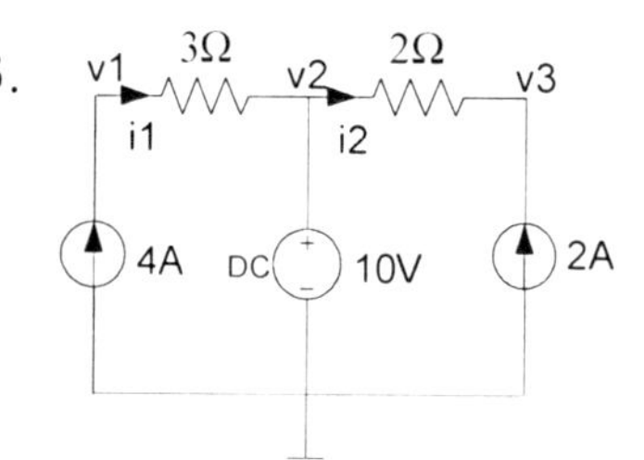

2.7   In the circuit to the right, find the two voltages, $v1$ and $v3$.

2.8   Find the equivalent resistance R_{AB}.

2.9   Find R_{AB} in the following circuit.

2.10   Find R_{AB} in the following circuit.

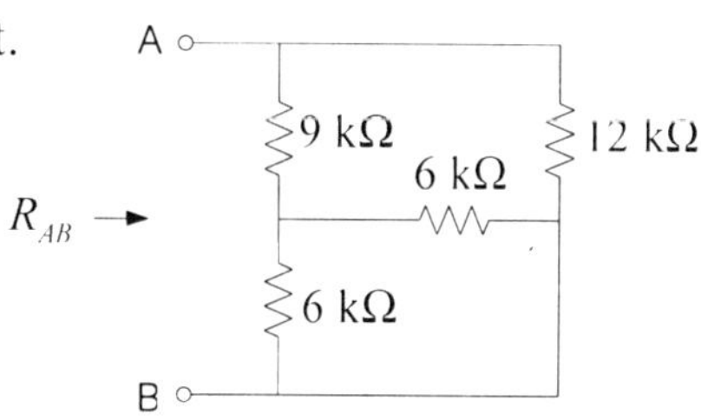

2.11 Find R_{AB} in the following circuit.

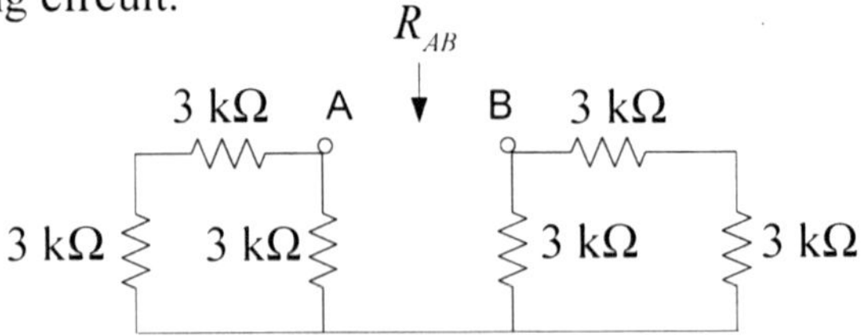

2.12 Suppose that the current source $i_0 = 2\sin\omega t$ is connected to a capacitor C as shown in the circuit to the right. If the frequency of the sine wave is 15 kHz and the inductance is 5 μF, use Matlab to plot the current i_C through the capacitor and the voltage v_C across the capacitor as a function of time for two cycles of the current sine wave.

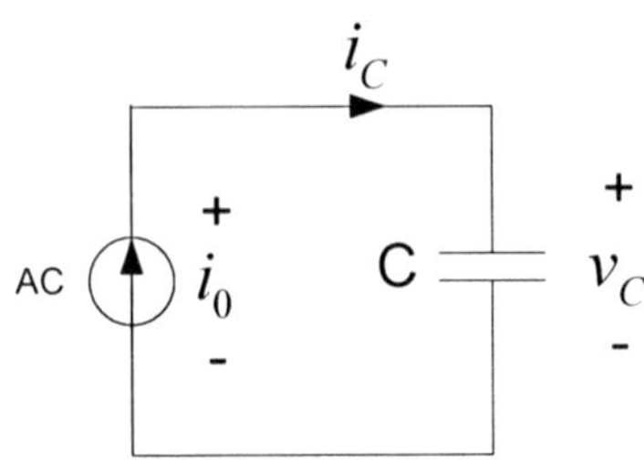

2.13 Suppose that the current source $i_0 = 2\sin\omega t$ is connected to an inductor L as shown in the circuit to the right. If the frequency of the sine wave is 2 kHz and the inductance is 1 mH, use Matlab to plot the current i_L through the inductor and the voltage v_L across the inductor as a function of time for two cycles of the current sine wave.

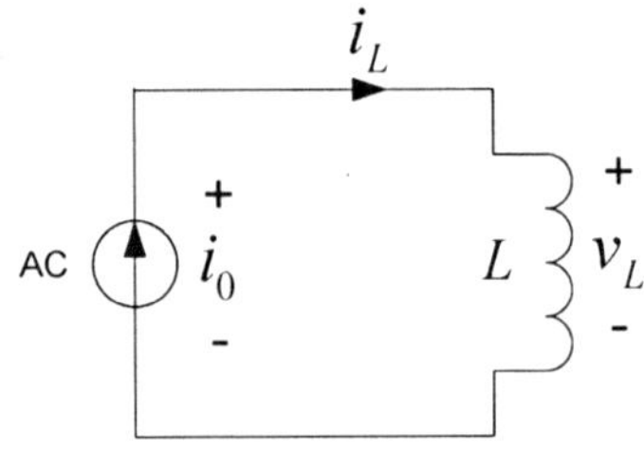

2.14 Find the equivalent capacitance C_{eq} between terminals a and b in the following circuit.

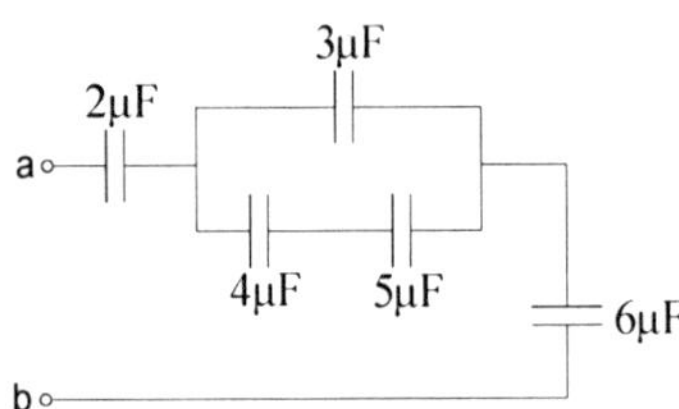

2.15 Find the equivalent inductance L_{eq} between terminals a and b in the following circuit.

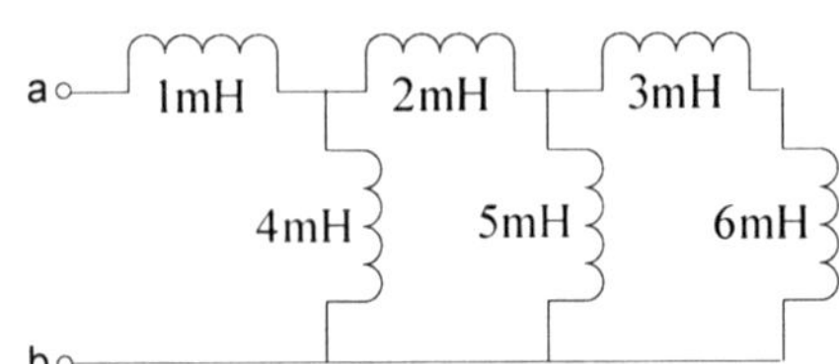

Chapter 3

Kirchhoff's Laws

3.1 Circuit Topology

The topology of a circuit describes how the circuit is laid out. Consider the circuit shown in Fig. 3.1. We will define a *branch* to be a single circuit element, i.e. any two-terminal element. In Fig. 3.1, there are five branches: the three resistors and the two voltage sources. A *node* is the point of connection between two or more branches. In Fig. 3.1, there are four nodes: *a*, *b*, *c*, and *d*. A *loop* is any closed path in the circuit. In Fig. 3.1, there are three loops: *abcda*, *abda*, and *bcdb*. A *mesh* is a loop that doesn't contain any other loops. In Fig. 3.1, there are two meshes: *abda*, and *bcdb*.

The problem in circuit theory is to find the currents in the branches and the voltages at the nodes. One of the nodes is taken as a reference node and the voltages at the other nodes are measured relative to this reference node. In Fig. 3.1 we take node *d* to be the reference node and give it an arbitrary voltage of 0V or ground. Note that we already know the voltages at nodes *a* and *c* because they are given by the independent voltage sources. The current in branches *da* and *ab* will be the same, and also the current in branches *bc* and *cd* will be the same.

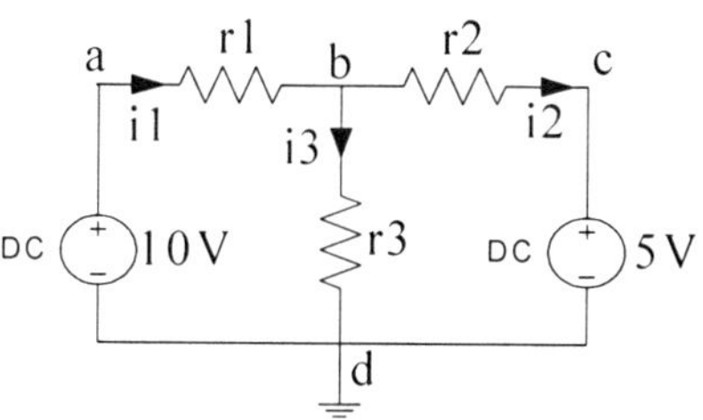

Figure 3.1 A sample circuit

We can represent the circuit in Fig. 3.1 by the directed graph shown in Fig. 3.2a. This graph has 5 branches and 4 nodes. Let's see if we can derive an expression for the number of independent loops in any graph.

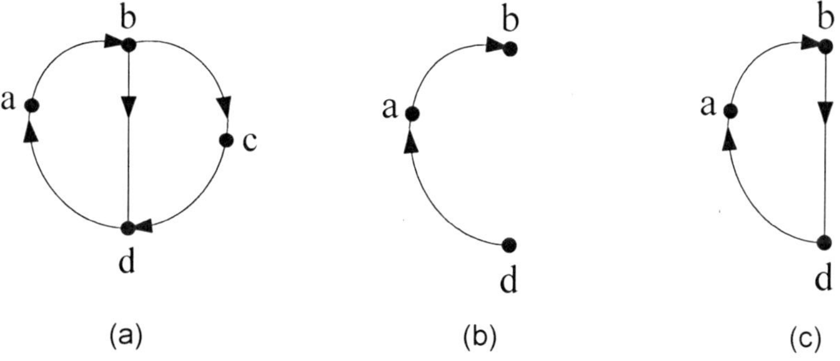

Figure 3.2 Directed graph representing the circuit in Fig. 3.1

Suppose we start drawing the graph shown in Fig. 3.2a. We could begin by drawing the branch *da*. We could then draw the branch *ab* as shown in Fig. 3.2b. At this point we have 2 branches and 3 nodes. Note that every time we add a branch we add a node. Counting the first node it is clear that the number of branches will be equal to the number of nodes minus 1. Now, if we draw the branch *bd*, we have added a new branch but not a new node. However, we have created the *loop dabd*. Thus, every time we connect a new branch to an existing node, we add a loop. Drawing the branch *bc* adds the new node *c* and drawing the branch *cd* adds a new loop. You could consider this new loop to be *dbcd* or *dabcd* but not both. When we add a branch to an existing node, we will create one new independent loop. Thus, we see that, in general, the number of branches *b* is given by

$$b = n + l - 1 \tag{3.1}$$

where *n* is the number of nodes and *l* is the number of independent loops. This is called the *Fundamental Theorem of Circuit Topology*. From Eq. (3.1), we can write the number of independent loops in the graph as

$$l = b - n + 1 \tag{3.2}$$

These are *independent* loops because each loop was generated by adding a new branch. In Fig. 3.2a, the number of branches *b* is 5 and the number of nodes *n* is 4. Therefore, $l = 5 - 4 + 1 = 2$.

What is nice about the Fundamental Theorem of Circuit Topology is that simply counting the number of branches and the number of nodes is sufficient to calculate the number of independent loops. We will see in Chapter 4 that *l* is also the number of equations we need to write if we use mesh analysis to calculate all currents and voltages in a circuit. It should be noted that loops *dabd* and *dbcd* are the two meshes in the circuit in Fig. 3.1 and that *meshes are always independent loops*.

3.2 Kirchhoff's Current Law

Recall from Section 1.3 that Kirchhoff's current law (KCL) states that *the algebraic sum of the currents entering (or leaving) a node (or a closed boundary) is zero.* That is,

$$\sum_{n=1}^{N} i_n = 0 \tag{3.3}$$

where *N* is the number of branches connected to the node and i_n is the nth current entering (leaving) the node. We will use the sign convention that currents entering a node are positive and currents leaving a node are negative. An equivalent statement of Kirchhoff's current law would be that *the sum of the currents entering a node is equal to the sum of the currents leaving the node.*

Example 8 – Kirchhoff's Current Law

In the circuit shown in Fig. 3.3, if $r1 = 6\Omega$, $r2 = 3\Omega$, and $r3 = 4\Omega$, find the voltage $v1$ and the three currents $i1$, $i2$, and $i3$.

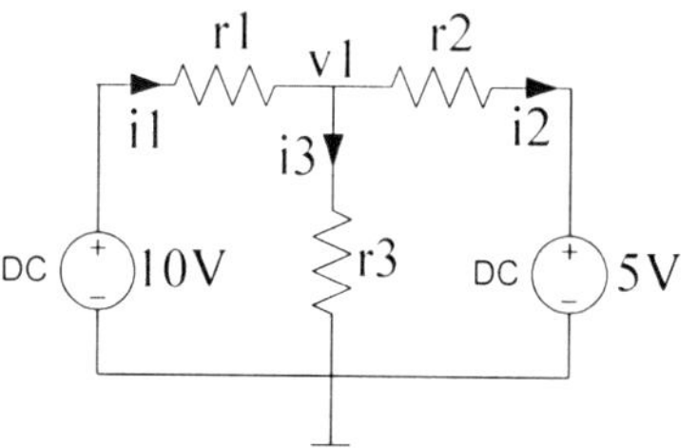

Figure 3.3 Example 8

If we knew $v1$, then we could use Ohm's law to find the three currents. But we can write each of the currents in terms of $v1$ and then use Kirchhoff's current law to solve for $v1$. Thus, summing the currents into the node $v1$ we can write

$$i1 - i2 - i3 = 0 \tag{3.4}$$

At this point, it is convenient to define the three conductances $g1 = 1/r1$, $g2 = 1/r2$, and $g3 = 1/r3$. Then from Ohm's law, the three currents are

$$i1 = (10 - v1)\,g1 \tag{3.5}$$
$$i2 = (v1 - 5)\,g2 \tag{3.6}$$
$$i3 = v1\,g3 \tag{3.7}$$

Substituting Eqs. (3.5) – (3.7) into Eq. (3.4), we can write

$$(10 - v1)\,g1 - (v1 - 5)\,g2 - v1\,g3 = 0$$

or

$$v1(g1 + g2 + g3) = 10g1 + 5g2$$

from which

$$v1 = \frac{10g1 + 5g2}{g1 + g2 + g3} \tag{3.8}$$

The Matlab solution is shown in Matlab Example 8. We see that $v1 = 4.4444\,\text{V}$, $i1 = 0.9259\,\text{A}$, $i2 = -0.1852\,\text{A}$, and $i3 = 1.1111\,\text{A}$.

Matlab Example 8

```
>> r1 = 6;
>> r2 = 3;
>> r3 = 4;
>> g1 = 1/r1;
>> g2 = 1/r2;
>> g3 = 1/r3;
>> v1 = (10*g1 + 5*g2)/(g1+g2+g3)
v1 =
     4.4444

>> i1 = (10 - v1)*g1
i1 =
    0.9259

>> i2 = (v1 - 5)*g2
i2 =
   -0.1852

>> i3 = v1*g3
i3 =
    1.1111
```

3.3 Kirchhoff's Voltage Law

Recall from Section 1.3 that Kirchhoff's voltage law (KVL) states that *the algebraic sum of the voltages around any loop is zero.* That is,

$$\sum_{m=1}^{M} v_m = 0 \qquad\qquad (3.9)$$

where M is the number of voltages in the loop and v_m is the mth voltage in the loop.

Example 9 – Kirchhoff's Voltage Law

Let's try to solve the same circuit as in Example 8 using Kirchhoff's voltage law. We have redrawn the circuit from Fig. 3.3 in Fig. 3.4 where we have labeled the voltages across each resistor. We have two independent loops (meshes) in this circuit so let's sum the voltages around each loop. The sum of the voltages around the left loop is

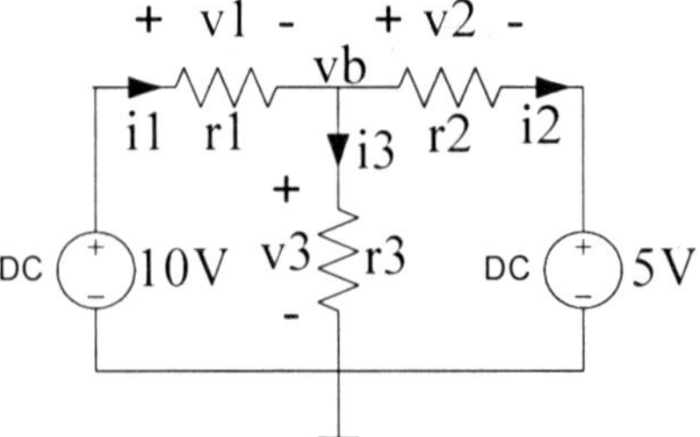

Figure 3.4 Example 9

$$-10V + v1 + v3 = 0$$

$$-10\text{V} + i1 \times r1 + i3 \times r3 = 0 \tag{3.10}$$

The sum of voltages around the right loop is

$$-v3 + v2 + 5\text{V} = 0$$

$$-i3 \times r3 + i2 \times r2 + 5\text{V} = 0 \tag{3.11}$$

Equations (3.10) and (3.11) are two equations in the three unknowns, $i1$, $i2$, and $i3$. We need another equation which we can get from Kirchhoff's current law given in Eq. (3.4).

$$i1 - i2 - i3 = 0 \tag{3.12}$$

Equations (3.10), (3.11), and (3.12) are three equations in three unknowns so we should be able to solve for the three currents $i1$, $i2$, and $i3$. Rearranging the terms we can write these three equations as

$$r1 \times i1 + 0 \times i2 + r3 \times i3 = 10\text{V}$$
$$0 \times i1 + r2 \times i2 - r3 \times i3 = -5\text{V}$$
$$1 \times i1 - 1 \times i2 - 1 \times i3 = 0$$

As shown in Appendix D we can rewrite these three equations in matrix form as

$$\begin{bmatrix} r1 & 0 & r3 \\ 0 & r2 & -r3 \\ 1 & -1 & -1 \end{bmatrix} \begin{bmatrix} i1 \\ i2 \\ i3 \end{bmatrix} = \begin{bmatrix} 10 \\ -5 \\ 0 \end{bmatrix}$$

or

$$\mathbf{Ri} = \mathbf{v} \tag{3.13}$$

Solving for the current vector $\mathbf{i}$

$$\mathbf{i} = \mathbf{R}^{-1}\mathbf{v} \tag{3.14}$$

A Matlab solution is shown in Matlab Example 9. Note that we write the 3x3 $\mathbf{R}$ matrix as

```
R = [r1 0 r3; 0 r2 -r3; 1 -1 -1]
```

where columns are separated by spaces (commas would work) and rows are separated by semi-colons. The column vector $\mathbf{v}$ is written as the transpose of a row vector using

```
v = [10 -5 0]'
```

You could have also written this column vector by using semi-colons after each element.

Matlab Example 9

```
>> r1 = 6;
>> r2 = 3;
>> r3 = 4;
>> R = [r1 0 r3; 0 r2 -r3; 1 -1 -1]
R =
     6       0       4
     0       3      -4
     1      -1      -1

>> v = [10 -5 0]'
v =
    10
    -5
     0

>> i = inv(R)*v
i =
    0.9259
   -0.1852
    1.1111

>> vb = i(3)*r3
vb =
    4.4444

>> i = R\v
i =
    0.9259
   -0.1852
    1.1111
```

Eq. (3.14) is written in Matlab as

$$i = inv(R)*v \tag{3.15}$$

which solves for the three currents. The node voltage *vb* is then computed using Ohm's law from

$$vb = i(3)*r3 \tag{3.16}$$

which agrees with Example 8. At the end of Matlab Example 9, we show that Eq. (3.12) can also be written in Matlab by using the left matrix divide operator \ by writing

$$i = R\backslash v \tag{3.17}$$

The backslash operator used for left division as in Eq. (3.17) is actually the preferred way of solving a system of equations in Matlab. The expression in Eq. (3.17) will work even if the matrix **R** is not square, whereas the expression *inv(R)* only works if **R** is a square matrix. The left divide operator computes a pseudo-inverse of **R** when **R** is not square. See Appendix D for more details about the left divide operator.

3.4 D/A Converter

In digital circuits the values of the signals are binary digits, or bits, that have values of either 0 or 1. A 0 might be represented by 0 volts and a 1 might be represented by 5 volts, or 3.3 volts, or 1 volt. A sequence of bits can represent binary numbers. A 3-bit binary number $b_2b_1b_0$ can represent the numbers 0 – 7 according to

$$N = b_2 \times 2^2 + b_1 \times 2^1 + b_0 \times 2^0 \tag{3.18}$$

These 8 binary numbers are shown in Table 3.1.

A digital-to-analog (D/A) converter (DAC) will convert a binary number to an analog voltage that is proportional to the corresponding decimal number. The R-2R ladder circuit shown in Fig. 3.6a will convert the 3-bit binary number $b_2b_1b_0$ to an analog voltage v_2 that is proportional to the decimal value of the binary number.

Table 3.1 Binary Numbers

Decimal number	Binary number
0	000
1	001
2	010
3	011
4	100
5	101
6	110
7	111

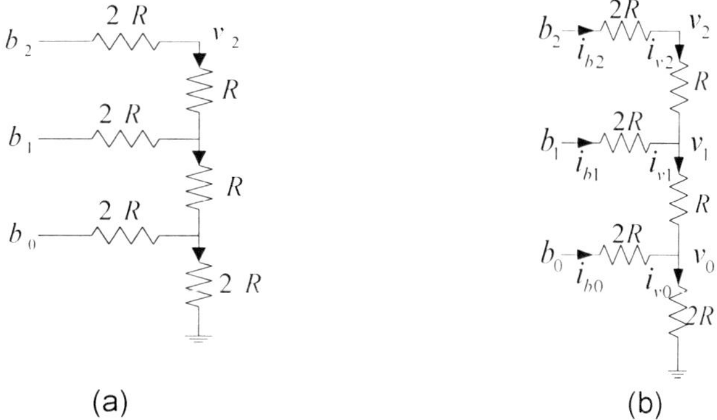

(a)   (b)

Figure 3.6 A R-2R Ladder DAC

To analyze this ladder network, we redraw the circuit in Fig. 3.6b where we have labeled the currents in each branch. We begin by applying Kirchhoff's current law to node v_0 and then using Ohm's law to write

$$i_{v0} = i_{b0} + i_{v1}$$

$$\frac{v_0}{2R} = \frac{b_0 - v_0}{2R} + \frac{v_1 - v_0}{R}$$

$$v_0 = b_0 - v_0 + 2v_1 - 2v_0$$

$$2v_0 = \frac{1}{2}b_0 + v_1 \qquad (3.19)$$

Now let's repeat the procedure for node v_1.

$$i_{v1} = i_{b1} + i_{v2}$$

$$\frac{v_1 - v_0}{R} = \frac{b_1 - v_1}{2R} + \frac{v_2 - v_1}{R}$$

$$2v_1 - 2v_0 = b_1 - v_1 + 2v_2 - 2v_1 \qquad (3.20)$$

If we now substitute Eq. (3.19) for the second term in Eq. (3.20), we obtain

$$2v_1 - \frac{1}{2}b_0 - v_1 = b_1 - v_1 + 2v_2 - 2v_1$$

$$2v_1 = \frac{1}{4}b_0 + \frac{1}{2}b_1 + v_2 \qquad (3.21)$$

Finally, using KCL and Ohm's law at node v_2, we can write

$$i_{v2} = i_{b2}$$

$$\frac{v_2 - v_1}{R} = \frac{b_2 - v_2}{2R}$$

$$2v_2 - 2v_1 = b_2 - v_2 \qquad (3.22)$$

If we substitute Eq. (3.21) for the second term in Eq. (3.22), we obtain

$$2v_2 - \frac{1}{4}b_0 - \frac{1}{2}b_1 - v_2 = b_2 - v_2$$

$$v_2 = \frac{1}{8}b_0 + \frac{1}{4}b_1 + \frac{1}{2}b_2 \qquad (3.23)$$

Eq. (3.23) gives the voltage v_2 as a function of the three input voltages b_0, b_1, and b_2. If these input voltages are either 0 or 1 volt representing the digital logic values 0 and 1, then the results of Eq. (3.23) are shown in Table 3.2. Note that the voltage v_2 increases linearly as the 3-bit binary number counts from 0 to 7. A plot of the data in Table 3.2 is shown in Fig. 3.7.

Table 3.2 Output of D/A converter

b_2	b_1	b_0	v_2
0	0	0	0
0	0	1	1/8
0	1	0	1/4
0	1	1	3/8
1	0	0	1/2
1	0	1	5/8
1	1	0	3/4
1	1	1	7/8

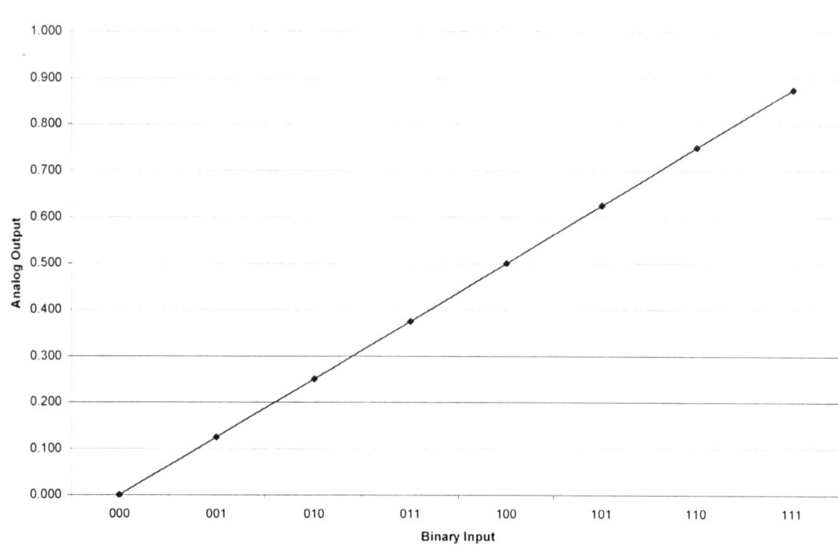

Figure 3.7 A plot of the data in Table 3.2

Example 10 – D/A Converter

Matlab Example 10 shows how to use Matlab to create a plot similar to Fig. 3.7. The resulting plot is shown in Fig. 3.8.

Matlab Example 10

```
>> b2 = [0 0 0 0 1 1 1 1]
b2 =
     0      0      0      0      1      1      1      1

>> b1 = [0 0 1 1 0 0 1 1]
b1 =
     0      0      1      1      0      0      1      1

>> b0 = [0 1 0 1 0 1 0 1]
b0 =
     0      1      0      1      0      1      0      1

>> v2 = b0/8 + b1/4 + b2/2
v2 =
  Columns 1 through 7
        0    0.1250    0.2500    0.3750    0.5000    0.6250    0.7500
  Column 8
    0.8750

>> t = 0:7
t =
     0      1      2      3      4      5      6      7

>> plot(t,V2)
```

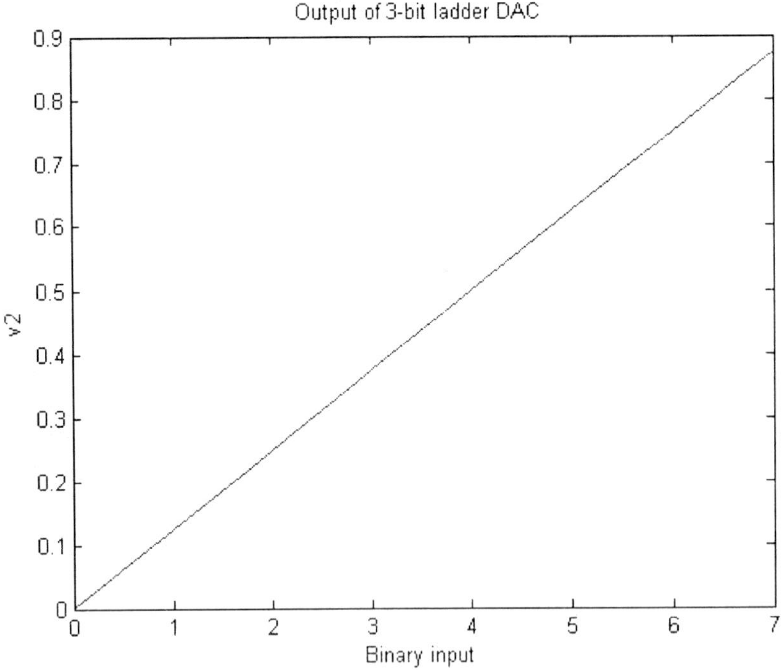

Figure 3.8 The plot generated from Matlab Example 10

Problems

3.1 In the circuit below, if $r1 = 4$ kΩ, $r2 = 8$ kΩ, and $r3 = 12$ kΩ, find the voltage $v1$ and the three currents, $i1$, $i2$, and $i3$.

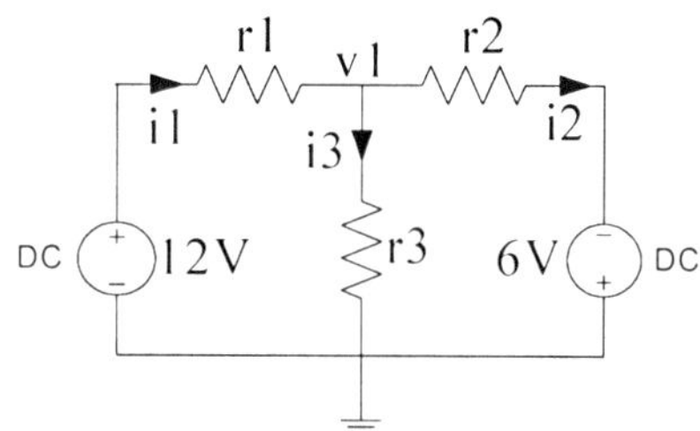

3.2 Repeat Problem 3.1 for values of $r1 = 1$ kΩ, $r2 = 1$ kΩ, and $r3 = 2$ kΩ.

3.3 In the following circuit, show that the voltage V is given by

$$V = \frac{1}{16} B_0 + \frac{1}{8} B_1 + \frac{1}{4} B_2 + \frac{1}{2} B_3$$

3.4 In the following circuit show that the voltage V_0 is given by

$$V_0 = \frac{4}{7} B_2 + \frac{2}{7} B_1 + \frac{1}{7} B_0$$

Chapter 4

Sinusoidal Signals and Phasors

In this chapter, we will show how sinusoidal signals can be represented by phasors. The use of phasors to represent sinusoidal voltages and currents in a circuit will allow us to define the impedances of capacitors and inductors as complex numbers. We will see that Ohm's law and Kirchhoff's laws hold for steady-state AC circuits if we simply replace resistances with complex impledances.

4.1 Sinusoidal Signals

A *sinusoid* is a signal that has the form of the sine or cosine function shown in Fig. 4.1.

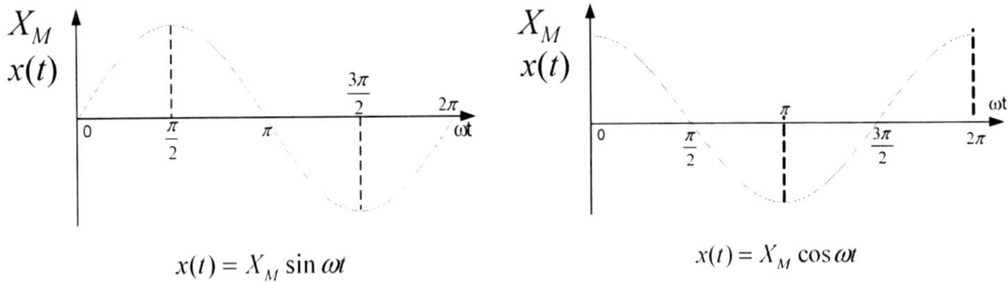

Figure 4.1 Sine and cosine waves

The angular frequency ω is equal to $2\pi f$ where the frequency f has the units hertz (Hz) and the period $T = 1/f$ is in seconds. Note that $\sin(\pi/2) = 1$ and $\cos(\pi/2) = 0$.

In general, sinusoids can have an initial phase angle ϕ as shown in Fig. 4.2.

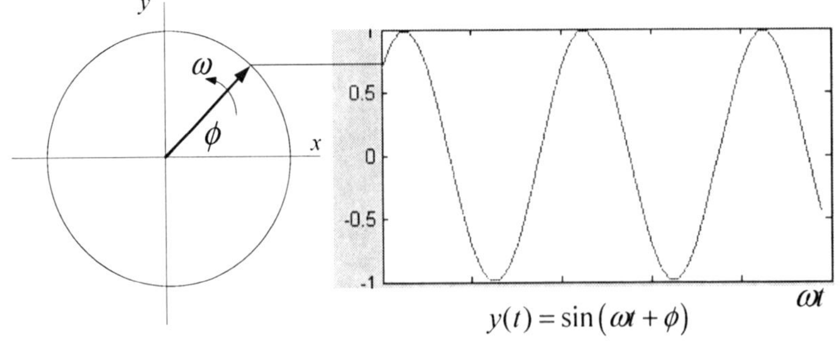

Figure 4.2 Defining the angle ϕ of a sine wave

Note that the sine wave in Fig. 4.2 can be visualized as the projection of a rotating unit vector on the y-axis that starts at an angle ϕ from the x-axis and rotates at the constant angular velocity ω. From this figure, you should be able to see that

$$\sin\left(\omega t + \frac{\pi}{2}\right) = \cos(\omega t) \qquad (4.1)$$

$$\sin(\omega t + \pi) = -\sin(\omega t) \qquad (4.2)$$

$$\sin\left(\omega t - \frac{\pi}{2}\right) = -\cos(\omega t) \qquad (4.3)$$

A sinusoidal voltage or current is usually referred to as an alternating current (or AC) voltage or current and circuits excited by sinusoids are called AC circuits. Sinusoids are important for several reasons:

1. Nature itself is characteristically sinusoidal (the pendulum, waves, etc.).
2. A sinusoidal current or voltage is easy to generate.
3. Through Fourier analysis, any practical periodic signal (one that repeats itself with a period T) can be represented by a sum of sinusoids.
4. A sinusoid is easy to handle mathematically; its derivative and integral are also sinusoids.
5. **When a sinusoidal source is applied to a linear circuit, the steady-state response *is also sinusoidal*, and we call the response the *sinusoidal steady-state response*.**

Example 11 – Sinusoidal Signals

Plot two cycles of $\sin(\omega t + \phi)$ for different values of ϕ.

To do this, it is convenient to define a Matlab *function* that we can call to plot $\sin(\omega t + \phi)$ for different values of ϕ. You do this by writing the function in a Matlab *.m* file. Make sure the *Current Directory* in Matlab is pointing to some convenient directory where you want to store you *.m* file. Then select

```
File -> New -> M-File
```

A Matlab Editor window will open. Type in the program shown in Listing 4.1 and save the file as *plotsin.m*. The first statement in the program

```
function plotsin(phi)
```

defines *plotsin* to be a Matlab function with a single parameter called *phi*. The next five lines are comment lines that begin with the % symbol. You should always include comments in your functions that define what the function does, define each parameter, and give the form in which the function is called. Note that the rest of the program is just a listing of Matlab statements that you could have typed directly into Matlab as we have been doing up until this point. The difference now is that these statements don't get executed until you call the function in Matlab. The way to do this is shown in Matlab Example 11.

Listing 4.1 plotsin.m

```
function plotsin(phi)
% plot sin(wt + phi) for 2 cycles
% phi = angle in radians
% wt goes from 0 to 4*pi
% plotsin(phi)
%
wt = linspace(0,4*pi,80);
y = sin(wt + phi);
plot(wt,y)
```

Matlab Example 11

```
>> help plotsin
  plot sin(wt + phi) for 2 cycles
  phi = angle in radians
  wt goes from 0 to 4*pi
  plotsin(phi)


>> plotsin(0)
>> plotsin(pi/4)
>> plotsin(pi/2)
>> plotsin(3*pi/4)
>> plotsin(pi)
>> plotsin(5*pi/4)
>> plotsin(3*pi/2)
>> plotsin(7*pi/4)
```

If you type

```
help plotsin
```

then all of the comments that are included in Listing 4.1 will be displayed on the screen. This will remind you how to call the function. Try all of the examples shown in Matlab Example 11 and see if you can predict how the plot will look based on Fig. 4.2. The result of typing

```
plotsin(5*pi/4)
```

is shown in Fig. 4.3

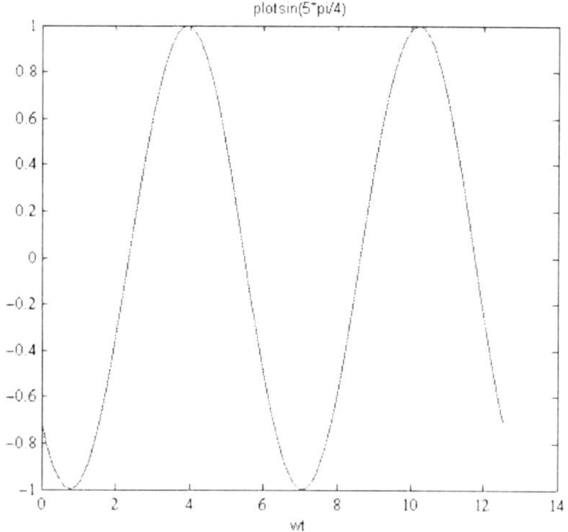

Figure 4.3 Plot resulting from $plotsin(5*pi/4)$

In the next section, we will see that it is convenient to consider the *x-y* plane in Fig. 4.2 to be the complex plane in which complex numbers are defined.

4.2 Root-Mean-Square (RMS) Values

If you measure the voltage coming out of a wall plug in your home with an oscilloscope, you will obtain a 60Hz sinusoidal wave that will look something like Fig. 4.4.

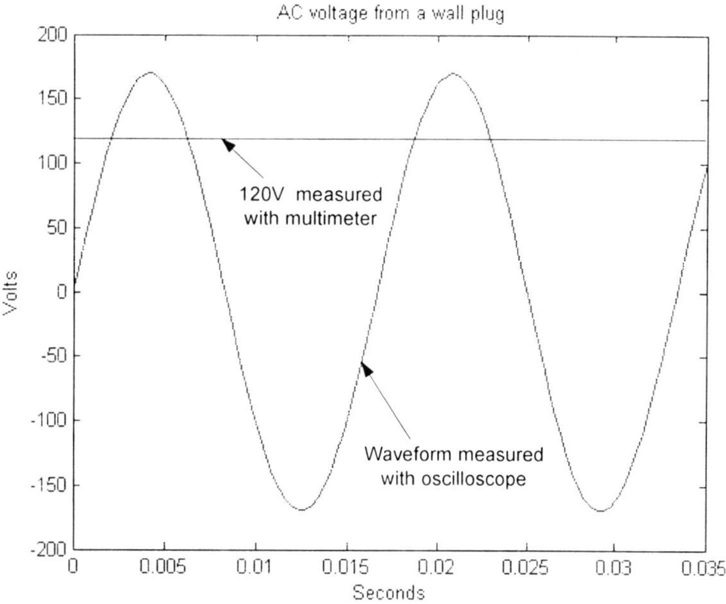

Figure 4.4 A 120V, 60Hz sine wave

On the other hand, if you measure the same voltage using a multimeter, it might record 120V as shown in Fig. 4.4. But what is this 120V? It certainly isn't the maximum value of the sine wave which is almost 170V, and it can't be the average value of the sine wave which is zero! It turns out to be what we call the RMS, or *root-mean-square*, value of the sine wave.

We define the *effective* or *rms* value of a periodic voltage (current) source to be the dc voltage (current) that delivers the same average power to a resistor. Thus, if $v(t)$ is a periodic waveform with period T, and if V_{rms} is a dc voltage source, the average power delivered to a resistor, R, by the voltage (or current) source during T seconds will be

$$P = \frac{V_{rms}^2}{R} = \frac{1}{T}\int_0^T \frac{v^2(t)}{R}dt \tag{4.4}$$

from which

$$V_{rms} = \sqrt{\frac{1}{T}\int_0^T v^2(t)dt} \tag{4.5}$$

In terms of the current through R, we would write

$$P = I_{rms}^2 R = \frac{1}{T}\int_0^T i^2(t)Rdt \tag{4.6}$$

from which

$$I_{rms} = \sqrt{\frac{1}{T}\int_0^T i^2(t)dt} \tag{4.7}$$

Thus, to find the *rms* value of any periodic waveform you first square the value of the waveform, then you take the average, or mean, value over a period, and then you take the square root of the result. It is therefore the *root* of the *mean* of the *square*, or root-mean-square, or
rms.

Let's calculate the rms value of the sinusoidal voltage

$$v(t) = V_M \cos(\omega t + \theta_v) \tag{4.8}$$

If we substitute Eq. (4.5) in Eq. (4.2) and use the trigonometric identity

$$\cos^2\phi = \tfrac{1}{2} + \tfrac{1}{2}\cos 2\phi \tag{4.9}$$

we obtain

$$V_{rms} = V_M\left[\frac{\omega}{2\pi}\int_0^{2\pi/\omega}\left[\tfrac{1}{2} + \tfrac{1}{2}\cos(2\omega t + 2\theta_v)\right]dt\right]^{1/2} \tag{4.10}$$

where we have used the fact that $T = 2\pi/\omega$. Now the integral of $\cos(2\omega t + 2\theta_v)$ over one period T will be zero because the area under the positive portion of the curve will exactly match the area under the negative portion of the curve. Therefore, Eq. (4.7) becomes

$$V_{rms} = V_M \left[\frac{\omega}{2\pi} \int_0^{2\pi/\omega} \frac{1}{2} dt \right]^{1/2} = V_M \left[\frac{\omega}{2\pi} \left(\frac{t}{2} \right) \Big|_0^{2\pi/\omega} \right]^{1/2} = \frac{V_M}{\sqrt{2}} \qquad (4.11)$$

Note that V_{rms} as given in Eq. (4.11) is independent of the phase angle θ_v in Eq. (4.8). In particular, it will hold for our sine wave in Fig. 4.1. From Eq. (4.11), we can write the maximum value of the sine wave as

$$V_M = V_{rms} \sqrt{2} \qquad (4.12)$$

If $V_{rms} = 120\text{V}$, then $V_M = 120\text{V} \sqrt{2} = 169.7\text{V}$ as shown in Fig. 4.1.

Example 12 – RMS Voltage

Calculate the rms value, V_{rms}, and the average value, V_{avg}, of the pulse-width modulated (PWM) signal, $v(t)$, shown in Fig. 4.5. A PWM signal has a constant period, T, and the high time of the square wave is some fraction α of the period called the *duty cycle*. Plot V_{rms} and V_{avg} as a function of α for values of α between 0 and 1.

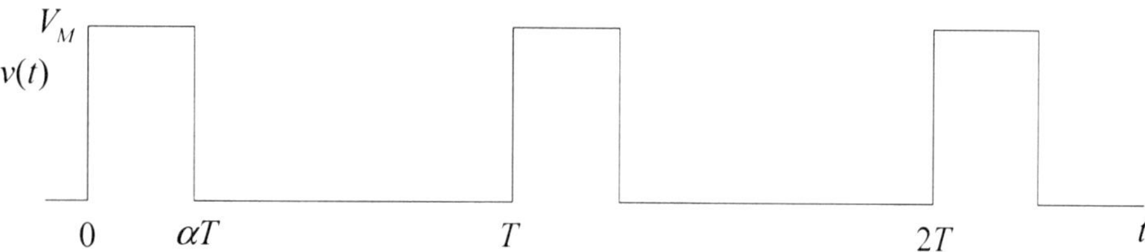

Figure 4.5 A pulse-width modulated (PWM) signal

From Eq. (4.5),

$$V_{rms} = \sqrt{\frac{1}{T} \int_0^T v^2(t) dt}$$

$$= \sqrt{\frac{1}{T} \left(\int_0^{\alpha T} V_M^2 dt + \int_{\alpha T}^T 0 dt \right)} = \sqrt{\frac{1}{T} V_M^2 \, t \Big|_0^{\alpha T}}$$

or

$$V_{rms} = V_M \sqrt{\alpha} \qquad (4.13)$$

The average value of the waveform in Fig. 4.5 is given by

$$V_{avg} = \frac{1}{T} \int_0^T v(t) dt \qquad (4.14)$$

$$= \frac{1}{T}\left(\int_0^{\alpha T} V_M \, dt + \int_{\alpha T}^{T} 0 \, dt \right) = \frac{V_M}{T} t \Big|_0^{\alpha T}$$

or

$$V_{avg} = \alpha V_M \qquad\qquad (4.15)$$

Matlab Example 12 is a Matlab program that will plot *Vrms* and *Vavg* as a function of α for $V_M = 1$V and values of α between 0 and 1. The results are shown in Fig. 4.6. Note that the rms value is always greater than the average value except at $\alpha = 0$ (no signal at all) or $\alpha = 1$ (a dc voltage and no waveform).

Matlab Example 12

```
>> Vm = 1;
>> a = 0:0.05:1;
>> Vrms = Vm*sqrt(a);
>> Vavg = a*Vm;
>> curves = [Vrms; Vavg];
>> plot(a,curves)
```

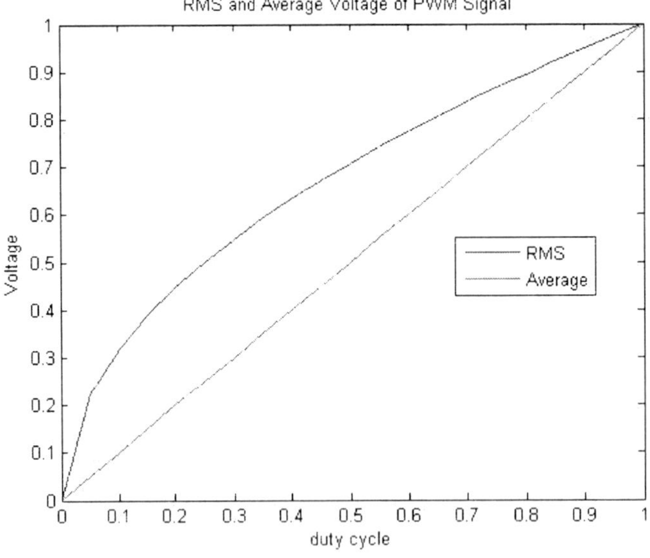

Figure 4.6 Plot resulting from Matlab Example 12

Example 13 – RMS Current

Calculate the rms value, I_{rms}, and the average value, I_{avg}, of the triangular current waveform, $i(t)$, shown in Fig. 4.7.

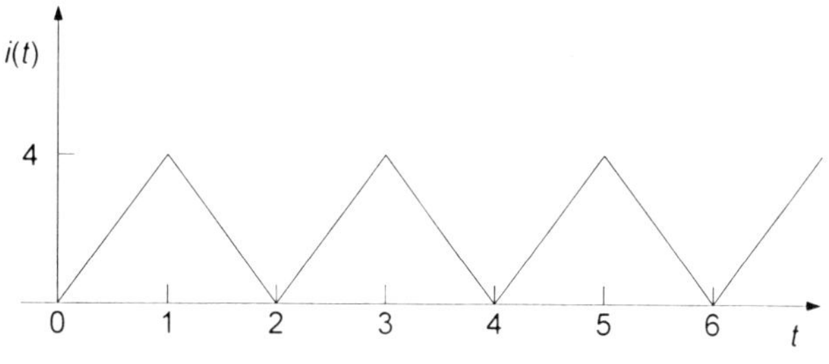

Figure 4.7 A triangle wave

For $0 < t < T$, the current waveform in Fig. 4.7 can be expressed as

$$i(t) = \begin{cases} 4t & 0 < t < 1 \\ 8 - 4t & 1 < t < 2 \end{cases} \qquad (4.16)$$

From Eq. (4.7),

$$I_{rms} = \sqrt{\frac{1}{T} \int_0^T i^2(t)dt}$$

$$= \sqrt{\frac{1}{2}\left(\int_0^1 (4t)^2 dt + \int_1^2 (8-4t)^2 dt \right)}$$

$$= \sqrt{0.5\left(\int_0^1 16t^2 dt + \int_1^2 \left(64 - 64t + 16t^2\right)dt \right)}$$

$$= \sqrt{0.5\left(\frac{16t^3}{3}\bigg|_0^1 + 64t\bigg|_1^2 - \frac{64t^2}{2}\bigg|_1^2 + \frac{16t^3}{3}\bigg|_1^2 \right)}$$

$$= \sqrt{0.5\left(\frac{16}{3} + 64(2-1) - 32(4-1) + \frac{16}{3}(8-1) \right)}$$

from which

$$I_{rms} = 2.309 \text{ A} \qquad (4.17)$$

as shown in Matlab Example 13. The average value of the waveform in Fig. 4.7 is given by

$$I_{avg} = \frac{1}{T} \int_0^T i(t)dt$$

$$= \frac{1}{2}\left(\int_0^1 4t\,dt + \int_1^2 (8-4t)dt \right)$$

$$= 0.5\left(\frac{4t^2}{2}\bigg|_0^1 + 8t\bigg|_1^2 - \frac{4t^2}{2}\bigg|_1^2 \right)$$

$$= 0.5\left(\frac{4}{2} + 8(2-1) - \frac{4}{2}(4-1) \right)$$

from which

$$I_{avg} = 2 \text{ A} \tag{4.18}$$

as shown in Matlab Example 13.

Matlab Example 13

```
>> Irms = sqrt(0.5*((16/3)+64*(2-1)-32*(4-1)+(16/3)*(8-1)))
Irms =
    2.3094

>> Iavg = 0.5*(2+8*(2-1)-2*(4-1))
Iavg =
    2
```

4.2 Complex Numbers and Phasors

The solution of the equation $x^2 = -1$ is $x = \pm\sqrt{-1}$. What is the meaning of $\sqrt{-1}$? It is an *imaginary* number that we call j.[5] Because $j = \sqrt{-1}$, it follows that $j^2 = -1$. It is therefore also true that

$$\frac{1}{j} = \frac{j}{jj} = \frac{j}{j^2} = -j \tag{4.19}$$

In general, numbers are complex and can have both a real and imaginary part. We represent such *complex numbers* in a *complex plane* in which the x-axis represents real numbers and the y-axis represents imaginary numbers. The unit circle in the complex plane is shown in Fig. 4.8. Points along the real axis in Fig. 4.8 represent *real numbers*; points along the imaginary axis in Fig. 4.8 represent *imaginary numbers*; and points in other parts of the complex plane in Fig. 4.8 represent *complex numbers*.

The complex number

$$\mathbf{A} = x + jy \tag{4.20}$$

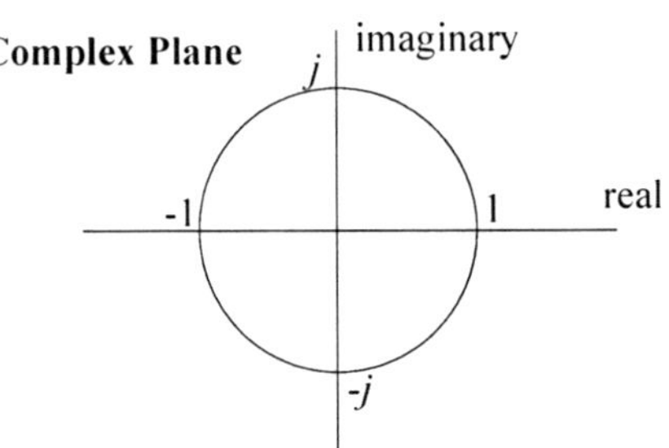

Figure 4.8 Unit circle in the complex plane

can be represented as a vector in the complex plane as shown in Fig. 4.9.

---

[5] Physicists and mathematicians use i for $\sqrt{-1}$. But electrical engineers already use i for current so for them, and you, $j = \sqrt{-1}$.

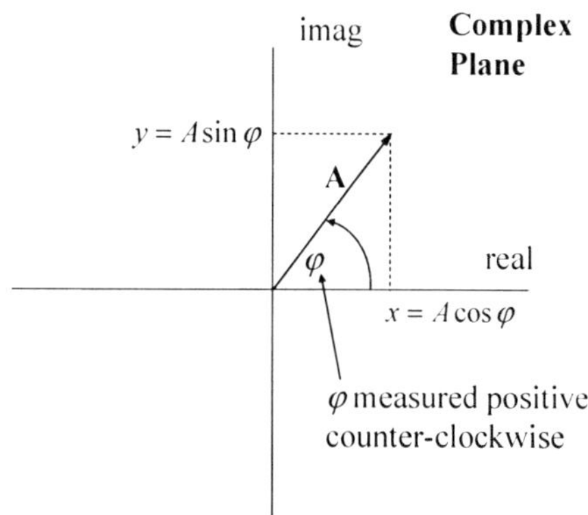

Figure 4.9 Complex numbers in the complex plane

From Fig. 4.9, we can write

$$\mathbf{A} = x + jy = A\cos\varphi + jA\sin\varphi \tag{4.21}$$

$$A = \sqrt{x^2 + y^2} \tag{4.22}$$

$$\varphi = \tan^{-1}\frac{y}{x} \tag{4.23}$$

Euler's equation is

$$e^{j\varphi} = \cos\varphi + j\sin\varphi \tag{4.24}$$

Therefore, we can write Eq. (4.21) as

$$\mathbf{A} = Ae^{j\varphi} \tag{4.25}$$

The vector $\mathbf{A}$ is called a *phasor*. Note that

$$e^{j(\theta+\phi)} = \cos(\theta+\phi) + j\sin(\theta+\phi) \tag{4.26}$$

If we let $\theta = \omega t$ in Eq. (4.26) then we can write

$$e^{j(\omega t+\phi)} = \cos(\omega t+\phi) + j\sin(\omega t+\phi) \tag{4.27}$$

Let's begin with $\phi = 0$. Then Eq. (4.27) becomes

$$e^{j\omega t} = \cos\omega t + j\sin\omega t \tag{4.28}$$

We see that $e^{j\omega t}$ is a complex number with a real part equal to $\cos\omega t$ and an imaginary part equal to $\sin\omega t$. Based on what you learned in Fig. 4.2, we see that $e^{j\omega t}$ can be represented by a rotating unit vector that plots out $\cos\omega t$ along the real axis and $\sin\omega t$ along the imaginary axis as shown in Fig. 4.10.

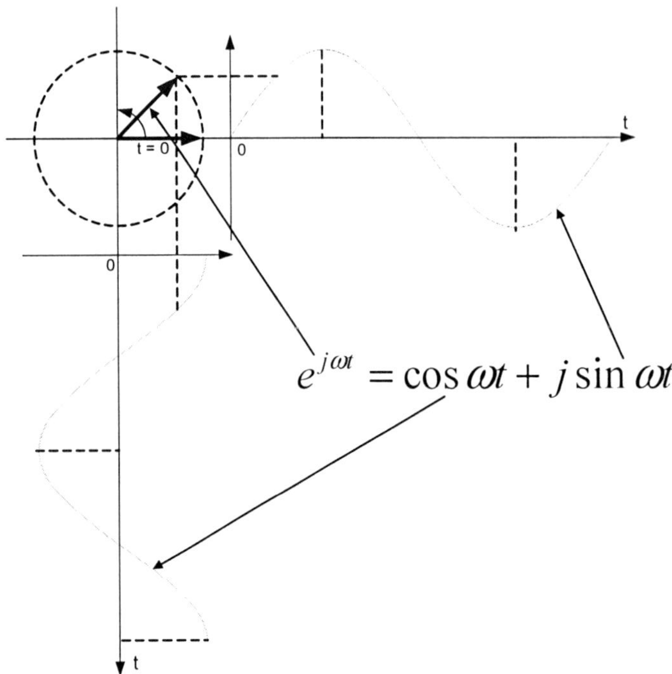

Figure 4.10 $e^{j\omega t}$ as a rotating unit phasor

We can represent any sinusoidal signal of the forms $\cos(\omega t + \phi)$ or $\sin(\omega t + \phi)$ as projections of rotating phasors on the *real* axis as shown in Fig. 4.11. Note that at $t = 0$ the unit phasor for $\cos\omega t$ will be along the positive real axis while the unit phasor for $\sin\omega t$ will be along the negative imaginary axis. In both cases, the proper waveform is projected on the real axis.

From Fig. 4.11, you can see that

$$\sin(\omega t + \phi) = \cos\left(\omega t - \left[\frac{\pi}{2} - \phi\right]\right) \tag{4.29}$$

or

$$\sin(\omega t + \phi) = \cos\left(\omega t + \phi - \frac{\pi}{2}\right) \tag{4.30}$$

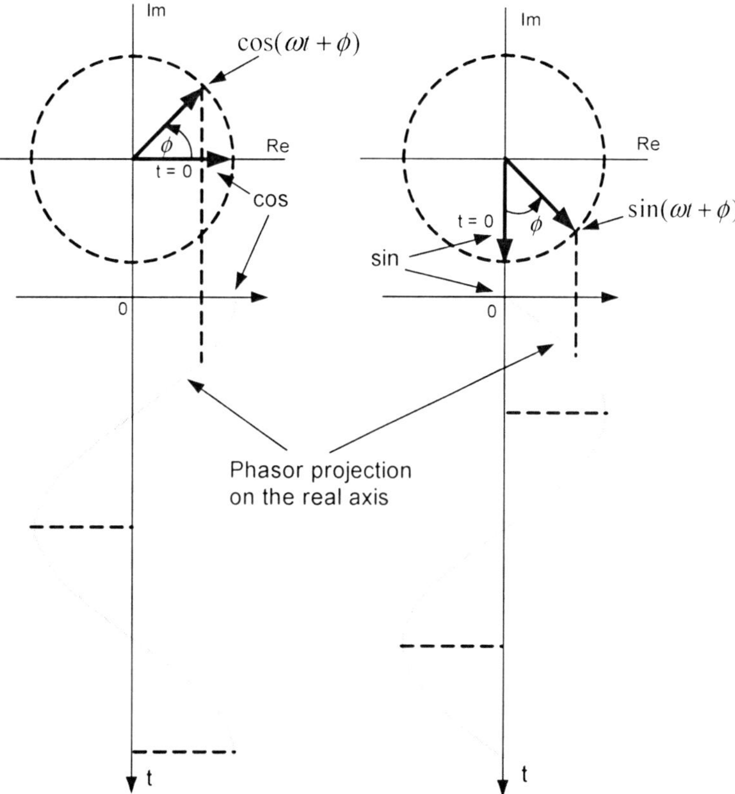

Figure 4.11 Relationship between *sin* and *cos*

Other relationships that you can verify by looking at Fig. 4.11 are

$$\sin\left(\omega t \pm \frac{\pi}{2}\right) = \pm\cos\omega t \tag{4.31}$$

$$\sin\left(\omega t \pm \pi\right) = -\sin\omega t \tag{4.32}$$

$$\cos\left(\omega t \pm \frac{\pi}{2}\right) = \mp\sin\omega t \tag{4.33}$$

$$\cos\left(\omega t \pm \pi\right) = -\cos\omega t \tag{4.34}$$

These relationships are summarized in Fig. 4.12a and can also be verified from the trigonometric identities

$$\sin\left(\omega t \pm \phi\right) = \sin\omega t\cos\phi \pm \cos\omega t\sin\phi \tag{4.35}$$

$$\cos\left(\omega t \pm \phi\right) = \cos\omega t\cos\phi \mp \sin\omega t\sin\phi \tag{4.36}$$

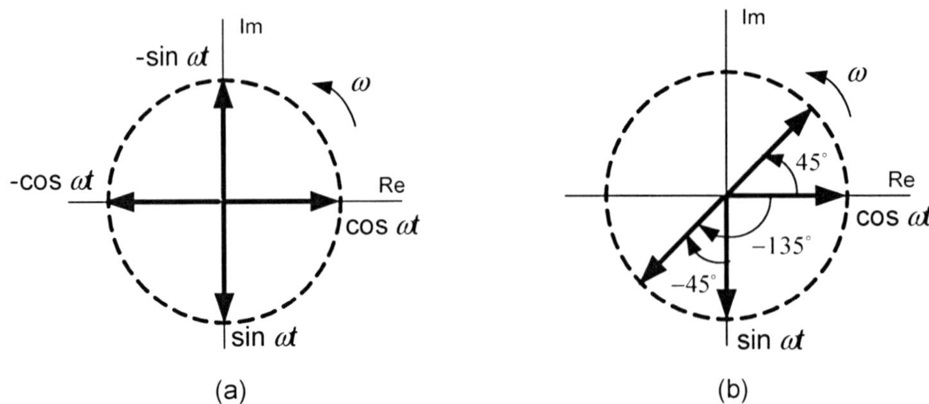

Figure 4.12 Examples of *sin* and *cos*

Note in Fig. 4.11 that the sine wave has the same form as the cosine wave, just delayed by $90°$. That is, the cosine wave starts at 1 at time $t = 0$, but the sine wave doesn't get to 1 until time $t = T/4$ or $90°$. We say that $\sin \omega t$ *lags* $\cos \omega t$ by $90°$. This can be seen in Fig. 4.12 where the $\sin \omega t$ phasor is $90°$ behind the $\cos \omega t$ phasor. We also say that $\cos \omega t$ *leads* $\sin \omega t$ by $90°$. In Fig. 4.12, positive angles are counter-clockwise, and

- $\cos \omega t$ *lags* $-\sin \omega t$ by $90°$
- $\sin\left(\omega t - 45°\right) = \cos\left(\omega t - 135°\right)$
- $\cos\left(\omega t + 45°\right)$ *leads* $\sin \omega t$ by $135°$

In our AC circuits, the independent voltage and current sources will be of the form $A\cos\left(\omega t + \phi\right)$ or $A\sin\left(\omega t + \phi\right)$. But from Eq. (4.27), we know that

$$Ae^{j(\omega t + \phi)} = A\cos\left(\omega t + \phi\right) + jA\sin\left(\omega t + \phi\right) \qquad (4.37)$$

Therefore, $A\cos\left(\omega t + \phi\right)$ is the real part of $Ae^{j(\omega t + \phi)}$ and $A\sin\left(\omega t + \phi\right)$ is the imaginary part of $Ae^{j(\omega t + \phi)}$. It turns out to be easier to use $Ae^{j(\omega t + \phi)}$ in our circuit equations than it is to use $A\cos\left(\omega t + \phi\right)$ because the derivative of $Ae^{j(\omega t + \phi)}$ with respect to t is just $j\omega Ae^{j(\omega t + \phi)}$. We will see in the next section that the $e^{j\omega t}$ part of $Ae^{j(\omega t + \phi)} = Ae^{j\omega t}e^{j\phi}$ gets cancelled out and we are left with just phasors of the form

$$\mathbf{A} = Ae^{j\phi} = A\angle\phi \qquad (4.38)$$

When these phasors are rotated at an angular speed of ω radians per second (corresponding to multiplying by $e^{j\omega t}$), then they plot a sinusoidal wave projected on the real axis as shown in Fig. 4.11. Thus, a *phasor is a complex number that represents the amplitude and phase of a sinusoid.* You can add, subtract, multiply, and divide phasors (and any complex number) as will be shown in Examples 14 – 15.

Example 14 – Complex Numbers

Given the complex number $z = 4 + j3$, find the magnitude and angle of this complex number.

Recall that a complex number $\mathbf{A}$ can be written as $\mathbf{A} = Ae^{j\varphi} = x + jy$ and the relationships among x, y, A, and φ are given by Eqs. (4.21) – (4.23). In Matlab, it is easy to find these relationships as shown in Matlab Example 14.

Matlab Example 14

```
>> z = 4 + 3j
z =
   4.0000 + 3.0000i

>> x = real(z)
x =
     4

>> y = imag(z)
y =
     3

>> mz = abs(z)
mz =
     5

>> az = angle(z)
az =
    0.6435

>> azd = angle(z)*180/pi
azd =
   36.8699

>> v = 5*exp(j*3*pi/4)
v =
   -3.5355 + 3.5355i

>> i = 8*exp(j*60*pi/180)
i =
   4.0000 + 6.9282i
```

Note that if we write

```
>> z = 4 + 3j
```

Matlab returns

```
z =
    4.0000 + 3.0000i
```

Thus, Matlab uses the physics and math value of $i = \sqrt{-1}$ rather than the j of electrical engineering. However, it will recognize $j = \sqrt{-1}$ if you use it in the proper context. Note that when writing complex numbers in Matlab we must write the complex number $z = 4 + j3$ as

```
>> z = 4 + 3j
```

with the j following the 3. Otherwise, Matlab will think that $j3$ is a variable name! Also note that

```
>> x = real(z)
```
and
```
>> y = imag(z)
```

can be used return the real and imaginary parts of z rather than having to use Eq. (4.21). To find the magnitude of z, we could use Eq. (4.22) and type

```
>> mz = sqrt(4^2+3^2)
```

which would work. However, it is easier to use the Matlab *abs* function and write

```
>> mz = abs(z)
```

as shown in Matlab Example 14. To find the angle of z, we could use Eq. (4.23) and type

```
>> az = atan(y/x)
```

which would work. However, it is easier to use the Matlab *angle* function and write
```
>> az = angle(z)
```

as shown in Matlab Example 14. Note that this angle, 0.6435, is given in radians. If you want the angle to come out in degrees, you would type

```
>> azd = angle(z)*180/pi
```

as shown in Matlab Example 14.

Suppose that you wanted to write the phasor voltage $\mathbf{V} = 5e^{j3\pi/4} = 5\angle 3\pi/4$ in Matlab. You would just type

```
>> v = 5*exp(j*3*pi/4)
```

as shown in Matlab Example 14. Note that Matlab returns the value

```
v =
    -3.5355 + 3.5355i
```

in rectangular coordinates so that you don't have to type

```
>> x = 5*cos(3*pi/4)
```

which will also give you a value of -3.5355. If you have a phasor current such as $\mathbf{I} = 8e^{j60^\circ} = 8\angle 60^\circ$ with the angle in degrees, then in Matlab you would need to type

```
>> i = 8*exp(j*60*pi/180)
```

which will return the value

```
i =
    4.0000 + 6.9282i
```

as shown in Matlab Example 14. If you want to find the real part of this current using the *cos* function, you must use the *cosd* function if the angle is in degrees. Thus,

```
>> x = 8*cosd(60)
```

will return 4.0000 without having to multiply the 60 by *pi*/180.

Example 15 – Complex Number Arithmetic

Given the two complex numbers $A = 6 + j2$ and $B = 3 + j5$, find the values of $C = A + B$, $D = A * B$, and $E = A/B$.

In general, it is easier to add (and subtract) complex numbers when they are in rectangular form because then you just add the real and imaginary parts. Thus, in this case,

$$C = A + B = (6+3) + j(2+5) = 9 + j7 \qquad (4.39)$$

We can multiply two complex numbers in either rectangular or polar forms. Thus,

$$C = A * B = (6 + j2)(3 + j5)$$
$$= (18 - 10) + j(6 + 30) = 8 + j36 \qquad (4.40)$$

Converting A and B to polar form,

$$A = 6 + j2 = 6.3246 e^{j0.322}$$

and

$$B = 3 + j5 = 5.831 e^{j1.030}$$

we could write

$$C = A * B = \left(6.3246e^{j0.322}\right)\left(5.831e^{j1.030}\right)$$
$$= (6.3246)(5.831)e^{j(0.322+1.030)} = 36.8782e^{j1.352} \qquad (4.41)$$

which is equivalent to Eq. (4.40). When dividing complex numbers "by hand", it is easiest to first convert to polar form. Thus,

$$C = \frac{6.3246e^{j0.322}}{5.831e^{j1.030}} = 1.085e^{j(0.322-1.030)} = 1.085e^{-j0.708} \qquad (4.42)$$

However, Matlab handles all arithmetic operations in both rectangular and polar form with equal ease as shown in Matlab Example 15.

Matlab Example 15

```
>> A = 6 + 2j
A =
   6.0000 + 2.0000i

>> B = 3 + 5j
B =
   3.0000 + 5.0000i

>> C = A + B
C =
   9.0000 + 7.0000i

>> D = A*B
D =
   8.0000 +36.0000i

>> E = A/B
E =
   0.8235 - 0.7059i

>> mA = abs(A)
mA =
   6.3246

>> mB = abs(B)
mB =
   5.8310

>> mD = abs(D)
mD =
   36.8782

>> mD = mA*mB
mD =
   36.8782
```

Matlab Example 15 (cont.)

```
>> aA = angle(A)
aA =
    0.3218

>> aB = angle(B)
aB =
    1.0304

>> aD = angle(D)
aD =
    1.3521

>> aD = aA + aB
aD =
    1.3521

>> mE = abs(E)
mE =
    1.0847

>> aE = angle(E)
aE =
    -0.7086

>> aE = aA - aB
aE =
    -0.7086

>> mE = mA/mB
mE =
    1.0847
```

4.3 Impedance and Admittance

In this section, we will see how the use of phasors makes the analysis of AC circuit almost as easy as the analysis of the DC circuits we studied in Chapters 2 – 5.

Capacitor

Let's begin with the capacitor circuit we used in Example 4 of Chapter 2 that is shown in Fig. 4.13. In Example 4, we assumed that the voltage v_0 was

$$v_0 = v_C = 5\sin(\omega t) \qquad (4.43)$$

where $\omega = 2\pi/T$ and the current i_C is given by

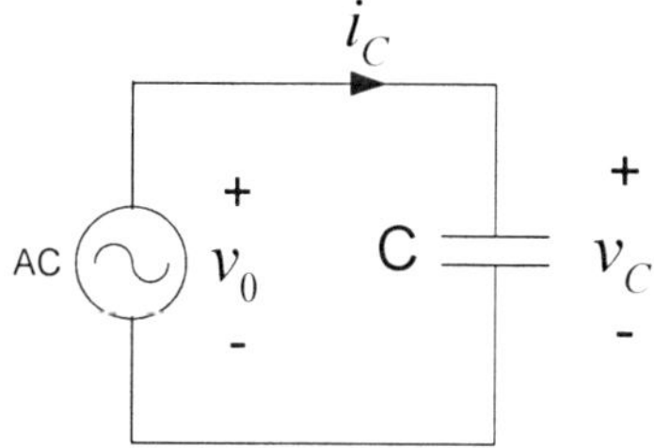

Figure 4.13 A simple capacitor circuit

$$i_C(t) = C \frac{dv_C(t)}{dt} \tag{4.44}$$

Now

$$5\sin(\omega t) = 5\cos\left(\omega t - \frac{\pi}{2}\right) = \text{Re}\left\{5e^{j\left(\omega t - \frac{\pi}{2}\right)}\right\} \tag{4.45}$$

Therefore, in Eq. (4.44), we will assume that

$$v_C = 5e^{j\left(\omega t - \frac{\pi}{2}\right)} \tag{4.46}$$

and then take the real part of the result to find the current i_C. Substituting Eq. (4.46) into Eq. (4.44) we obtain

$$i_C(t) = 5C \frac{d}{dt}\left(e^{j\left(\omega t - \frac{\pi}{2}\right)}\right) = 5Cj\omega e^{j\left(\omega t - \frac{\pi}{2}\right)}$$

$$= 5Cj\omega e^{-j\frac{\pi}{2}}e^{j\omega t} = 5Cj\omega(-j)e^{j\omega t} = 5C\omega e^{j\omega t} \tag{4.47}$$

Taking the real part of Eq. (4.47), we obtain

$$i_C(t) = 5C\omega\cos\omega t \tag{4.48}$$

which agrees with Eq. (2.25) in Chapter 2.
    Now Eq. (4.46) can be written as

$$v_C = 5e^{-j\frac{\pi}{2}}e^{j\omega t} = \mathbf{V}_C e^{j\omega t} \tag{4.49}$$

where

$$\mathbf{V}_C = 5e^{-j\frac{\pi}{2}} = 5\angle -90° \text{ V} \tag{4.50}$$

is a phasor voltage. Similarly, Eq. (4.47) can be written as

$$i_C = 5C\omega e^{j\omega t} = \mathbf{I}_C e^{j\omega t} \tag{4.51}$$

where

$$\mathbf{I}_C = 5C\omega e^{-j0} = 5C\omega\angle 0° \tag{4.52}$$

is a phasor current. Note that the phasor current in Eq. (4.52) leads the phasor voltage in Eq. (4.50) by 90 degrees as we found in Example 26. If you rotate these two phasors and project them on the real axis as we did it Fig. 4.7, you will obtain the same waveforms that we found in Fig. 2.18 in Example 4.

The *impedance* **Z** is defined as the phasor voltage divided by the phasor current and has the units of *ohms*. Thus, from Eqs. (4.35) and (4.37), we see that the impedance of the capacitor is given by

$$\mathbf{Z}_C = \frac{\mathbf{V}_C}{\mathbf{I}_C} = \frac{5e^{-j\frac{\pi}{2}}}{5C\omega e^{-j0}} = \frac{-j}{\omega C} = \frac{1}{j\omega C} \tag{4.53}$$

This impedance of the capacitor is purely imaginary and is called a *reactance*. The reciprocal of the impedance is called the *admittance*, **Y**, and has the units of *siemens*. Thus, from Eq. (4.53), the admittance of the capacitor is given by

$$\mathbf{Y}_C = \frac{\mathbf{I}_C}{\mathbf{V}_C} = j\omega C \tag{4.54}$$

This admittance of the capacitor is purely imaginary and is called a *susceptance*.

Inductor

We can do the same analysis of the inductor circuit we used in Example 6 of Chapter 2 that is shown in Fig. 4.14. In Example 6, we assumed that the voltage v_0 was

$$v_0 = v_L = 5\sin(\omega t) \tag{4.55}$$

where $\omega = 2\pi/T$ and the current i_L is given by

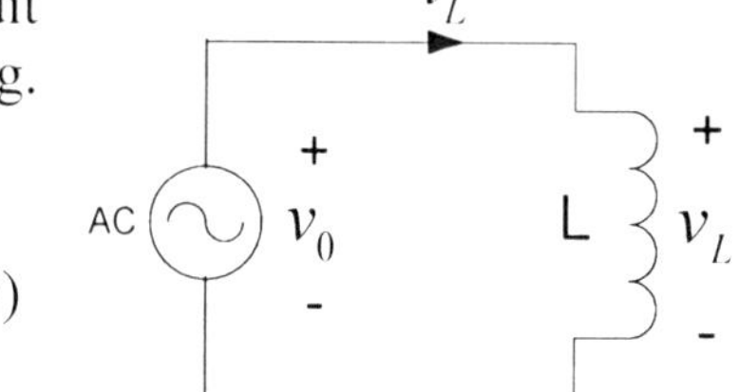

Figure 4.14 A simple inductor circuit

$$i_L(t) = i(t_0) + \frac{1}{L}\int_{t_0}^{t} v_L(t)dt$$

However, instead of integrating the voltage across the inductor to get the current as we did in Example 28, let's go back to the definition of voltage across the inductor from Eq. (7.22) and write

$$v_L(t) = L\frac{di_L(t)}{dt} \tag{4.56}$$

Using the same arguments we used to get Eq. (4.31), we will write the voltage across the inductor as

$$v_L = 5e^{j\left(\omega t - \frac{\pi}{2}\right)} = \mathbf{V}_L e^{j\omega t} \tag{4.57}$$

where

$$\mathbf{V}_L = 5e^{-j\frac{\pi}{2}} = 5\angle -90° \text{ V} \tag{4.58}$$

is a phasor voltage. Now we don't know the current $i_L(t)$ but we can assume that it is sinusoidal and we will therefore write

$$i_L = \mathbf{I}_L e^{j\omega t} \tag{4.59}$$

where $\mathbf{I}_L$ is an unknown phasor current to be determined. If we substitute Eqs. (4.57) and (4.59) into Eq. (4.56), we obtain

$$\mathbf{V}_L e^{j\omega t} = L\frac{d}{dt}\left(\mathbf{I}_L e^{j\omega t}\right) = j\omega L \mathbf{I}_L e^{j\omega t} \tag{4.60}$$

We can divide both sides of Eq. (4.60) by $e^{j\omega t}$ and obtain

$$\mathbf{V}_L = j\omega L \mathbf{I}_L \tag{4.61}$$

from which

$$\mathbf{I}_L = \frac{1}{j\omega L}\mathbf{V}_L \tag{4.62}$$

Substituting Eq. (4.58) into Eq. (4.62), we see that for this problem

$$\mathbf{I}_L = \frac{5e^{-j\frac{\pi}{2}}}{j\omega L} = \frac{-j5}{j\omega L} = -\frac{5}{\omega L} = \frac{5}{\omega L}e^{-j\pi} \tag{4.63}$$

Note that the phasor voltage in Eq. (4.58) leads the phasor current in Eq. (4.63) by 90 degrees as we found in Example 6. If you rotate these two phasors and project them on the real axis as we did it Fig. 4.10, you will obtain the same waveforms that we found in Fig. 2.27 in Example 6.

From Eq. (4.61), we see that the impedance of an inductor is

$$\mathbf{Z}_L = \frac{\mathbf{V}_L}{\mathbf{I}_L} = j\omega L \tag{4.64}$$

Note that the impedance of the inductor is also purely imaginary and is therefore a *reactance*. From Eq. (4.64), we see that the admittance of the inductor is given by

$$\mathbf{Y}_L = \frac{\mathbf{I}_L}{\mathbf{V}_L} = \frac{1}{j\omega L} \tag{4.65}$$

This admittance of the inductor is purely imaginary and is therefore a *susceptance*.

Example 16 – R-L Series Circuit

In the circuit shown in Fig. 4.15, find the current $i(t)$, the phasor voltages $\mathbf{V}_R$ and $\mathbf{V}_L$, and the impedance $\mathbf{Z}$ of the resistor and inductor in series. If $R = 1\text{k}\Omega$, $L = 1\text{mH}$, and $V_M = 2\text{V}$, plot the magnitudes of the phasor voltages $\mathbf{V}_R$ and $\mathbf{V}_L$ as a function of frequency from 10 kHz to 1 MHz.

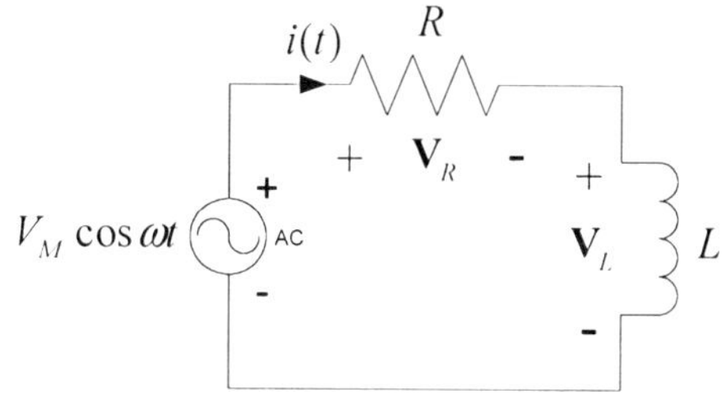

Figure 4.15 A series R-L circuit

We will write the voltage source as $\mathbf{V}e^{j\omega t}$ where

$$\mathbf{V} = V_M \angle 0^\circ \tag{4.66}$$

is the phasor voltage of the voltage source. Using KVL we can write

$$V_M e^{j\omega t} = Ri(t) + L\frac{di(t)}{dt} \tag{4.67}$$

Assume a sinusoidal current of the form

$$i(t) = I_M e^{j(\omega t + \phi)} = \mathbf{I}e^{j\omega t} \tag{4.68}$$

Substituting Eqs. (4.50) and (4.52) into Eq. (4.51), we can write

$$\mathbf{V}e^{j\omega t} = R\mathbf{I}e^{j\omega t} + j\omega L\mathbf{I}e^{j\omega t} \tag{4.69}$$

Dividing Eq. (4.53) by $e^{j\omega t}$ and solving for $\mathbf{I}$, we obtain

$$\mathbf{I} = \frac{\mathbf{V}}{R + j\omega L} = I_M \angle \phi \tag{4.70}$$

Recall that

$$R + j\omega L = \sqrt{R^2 + \omega^2 L^2}\, e^{j\tan^{-1}\left(\frac{\omega L}{R}\right)} \tag{4.71}$$

Therefore, substituting Eqs. (4.66) and (4.71) into Eq. (4.70), we obtain

$$\mathbf{I} = I_M \angle \phi = \frac{V_M}{\sqrt{R^2 + \omega^2 L^2}} \angle \left[-\tan^{-1}\frac{\omega L}{R} \right] \tag{4.72}$$

Substituting Eq. (4.72) into Eq. (4.68) and taking the real part, we obtain the current

$$i(t) = \frac{V_M}{\sqrt{R^2 + \omega^2 L^2}} \cos\left(\omega t - \tan^{-1} \frac{\omega L}{R}\right) \qquad (4.73)$$

To obtain the phasor voltages $\mathbf{V}_R$ and $\mathbf{V}_L$, note that Eq. (4.69) can be written as

$$\mathbf{V}e^{j\omega t} = \mathbf{V}_R e^{j\omega t} + \mathbf{V}_L e^{j\omega t} \qquad (4.74)$$

where

$$\mathbf{V}_R = R\mathbf{I} \qquad (4.75)$$

and

$$\mathbf{V}_L = j\omega L\mathbf{I} \qquad (4.76)$$

are the phasor voltages across the resistor and inductor. Note that, in each case, the phasor voltage is just the phasor current times the impedance of the element. Thus, we see that Ohm's law holds for AC circuits if we just use phasor voltages and currents and use impedance in place of resistance. Substituting the phasor current from Eq. (4.72) into Eqs. (4.75) and (4.76) would give the complete phasor voltages for the resistor and inductor. Dividing Eq. (4.74) by $e^{j\omega t}$ shows that Kirchhoff's voltage law (KVL) holds for AC circuits as long as you use the phasor voltages.

The total impedance $\mathbf{Z}$ of the resistor and inductor in series can be found directly from Eq. (4.70) as

$$\mathbf{Z} = \frac{\mathbf{V}}{\mathbf{I}} = R + j\omega L = Z\angle\theta_z \qquad (4.77)$$

where

$$Z = \sqrt{R^2 + \omega^2 L^2} \qquad (4.78)$$

is the magnitude of the impedance and

$$\theta_z = \tan^{-1} \frac{\omega L}{R} \qquad (4.79)$$

is the phase angle. Note that the total impedance of the two elements in series is just the sum of the impedances of each element. Thus, impedances in series add just like resistors in series. In general, the impedance is a *complex number*

$$\mathbf{Z} = R + jX \qquad (4.80)$$

containing a real, or *resistive* component R, and an imaginary, or *reactive*, component X.

Listing 4.2 is a Matlab program that plots the magnitudes of the phasor voltages $\mathbf{V}_R$ and $\mathbf{V}_L$ as a function of frequency from 10 kHz to 1 MHz. Matlab Example 16 runs the program for values of $R = 1$ kΩ, $L = 1$ mH, and $V_M = 2$ V. The resulting plot is shown in Fig. 4.11. In Listing 4.2, we just used Eq. (4.54) to calculate the phasor current at 100 different frequencies between 10 kHz and 1 MHz. Note from the plot in Fig. 4.16, that the voltage across the inductor goes up with frequency as expected. The voltage across the resistor must therefore go down because the sum of these two voltages must equal the input voltage of 2V. But note from the plot that the magnitudes of the two voltages don't add up to 2V. Remember that it is the *phasor* voltages that must be added together and the current through the inductor is 90 degrees out of phase with the voltage across the inductor. The current will be in phase with the voltage across the resistor, so the two phasor voltages plotted in Fig. 4.11 will be 90 degrees out of phase. Thus, when the magnitudes of the two voltages are equal, they should have a value of

$$V_R = V_L = 2\text{V}/\sqrt{2} = 1.414\text{V}$$

as shown in Fig. 4.17 and verified in Fig. 4.16. Note how KVL holds in Fig. 4.17.

Listing 4.2 RL.m

```
function RL(R,L,V)
% The voltage V is across R and L in series
% plot magnitude of Vr and Vl as function of frequency
%      from 10kHz to 1MHZ
% R = resistance in ohms
% L = inductance in henrys
% V = voltage in volts
% RL(R,L,V)
f = linspace(10000,1000000,100);
w = 2*pi.*f;
I = V./(R + w.*L*j);
Vr = R.*I;
Vl = w.*L*j.*I;
mVr = abs(Vr);
mVl = abs(Vl);
curves = [mVr; mVl];
plot(f,curves)
```

Matlab Example 16

```
>> help RL
  The voltage V is across R and L in series
  plot magnitude of Vr and Vl as function of frequency
       from 10kHz to 1MHZ
  R = resistance in ohms
  L = inductance in henrys
  V = voltage in volts
  RL(R,L,V)

>> R = 1000;
>> L = 0.001;
>> V = 2;
>> RL(R,L,V)
```

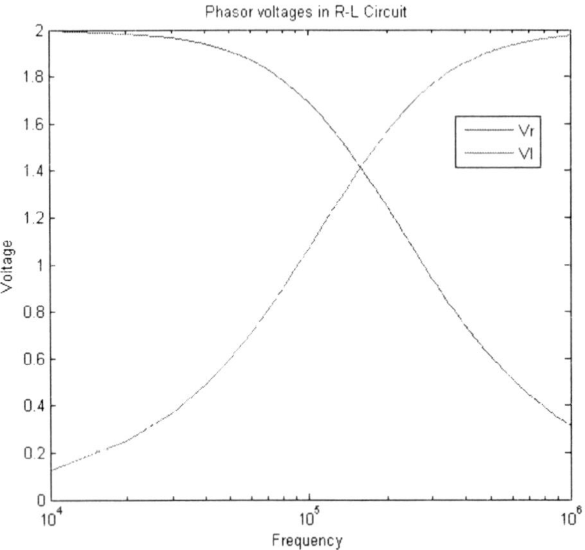

Figure 4.16 Plot resulting from Matlab Example 16

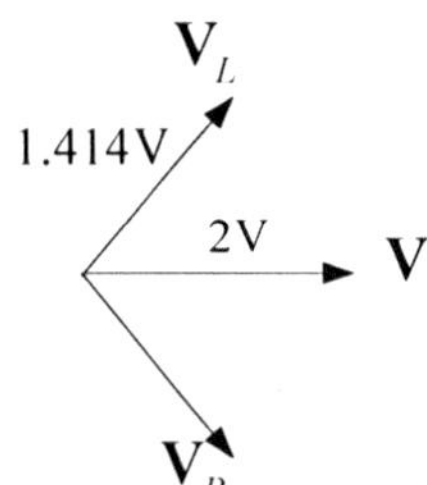

Figure 4.17 The phasor voltages when $V_R = V_L$ in Fig. 4.15

Example 17 – R-C Parallel Circuit

In the circuit shown in Fig. 4.18, find the voltage $v(t)$, the phasor currents $\mathbf{I}_R$ and $\mathbf{I}_L$, and the impedance $\mathbf{Z}$ of the resistor and capacitor in parallel. If $R = 470\ \Omega$, $C = 3$ nF, and $I_M = 2$ mA, plot the magnitudes of the phasor currents $\mathbf{I}_R$ and $\mathbf{I}_L$ as a function of frequency from 10 kHz to 1 MHz.

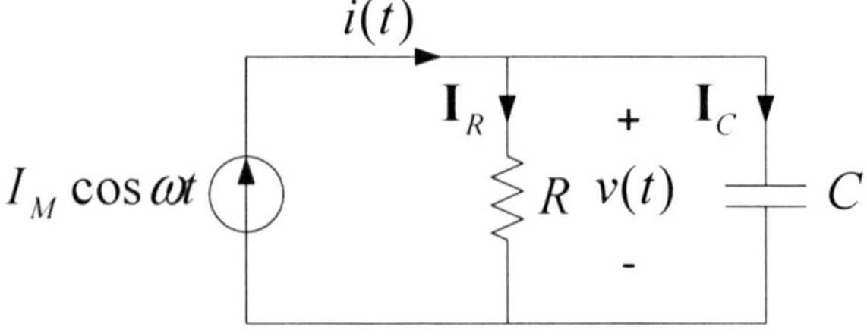

Figure 4.18 A parallel R-C circuit

We will write the current source as $\mathbf{I}e^{j\omega t}$ where

$$\mathbf{I} = I_M \angle 0^\circ \qquad\qquad (4.81)$$

is the phasor current of the current source. Using KCL, we can write

$$I_M e^{j\omega t} = \frac{v(t)}{R} + C\frac{dv(t)}{dt} \tag{4.82}$$

Assume a sinusoidal voltage of the form

$$v(t) = V_M e^{j(\omega t + \phi)} = \mathbf{V} e^{j\omega t} \tag{4.83}$$

Substituting Eqs. (4.81) and (4.83) into Eq. (4.82), we can write

$$\mathbf{I}e^{j\omega t} = \frac{\mathbf{V}e^{j\omega t}}{R} + j\omega C \mathbf{V} e^{j\omega t} \tag{4.84}$$

Dividing Eq. (4.84) by $e^{j\omega t}$ and solving for $\mathbf{V}$, we obtain

$$\mathbf{V} = \frac{\mathbf{I}}{\dfrac{1}{R} + j\omega C} = V_M \angle \phi \tag{4.85}$$

Therefore, substituting Eq. (4.81) into Eq. (4.85), we obtain

$$\mathbf{V} = V_M \angle \phi = \frac{I_M R}{\sqrt{1 + \omega^2 R^2 C^2}} \angle \left[-\tan^{-1}(\omega RC) \right] \tag{4.86}$$

Substituting Eq. (4.86) into Eq. (4.83) and taking the real part, we obtain the voltage

$$v(t) = \frac{I_M R}{\sqrt{1 + \omega^2 R^2 C^2}} \cos\left(\omega t - \tan^{-1}(\omega RC) \right) \tag{4.87}$$

To obtain the phasor currents $\mathbf{I}_R$ and $\mathbf{I}_C$, note that Eq. (4.84) can be written as

$$\mathbf{I}e^{j\omega t} = \mathbf{I}_R e^{j\omega t} + \mathbf{I}_C e^{j\omega t} \tag{4.88}$$

where

$$\mathbf{I}_R = \frac{\mathbf{V}}{R} \tag{4.89}$$

and

$$\mathbf{I}_C = j\omega C \mathbf{V} \tag{4.90}$$

are the phasor currents through the resistor and capacitor. Note that, in each case, the phasor current is just the phasor voltage divided by the impedance of the element. (Recall from Eq. (4.53) that the impedance of a capacitor is $1/j\omega C$.)

Thus, we see again that Ohm's law holds for AC circuits if we just use phasor voltages and currents and use impedance in place of resistance. Substituting the phasor voltage from Eq. (4.86) into Eqs. (4.89) and (4.90) would give the complete phasor currents for the resistor and capacitor. Dividing Eq. (4.88) by $e^{j\omega t}$ shows that Kirchhoff's current law (KCL) holds for AC circuits as long as you use the phasor currents.

The total impedance $\mathbf{Z}$ of the resistor and capacitor in parallel can be found directly from Eq. (4.85) as

$$\mathbf{Z} = \frac{\mathbf{V}}{\mathbf{I}} = \frac{1}{\dfrac{1}{R} + j\omega C} \tag{4.91}$$

or the admittance $\mathbf{Y}$ is

$$\mathbf{Y} = \frac{1}{\mathbf{Z}} = \frac{1}{R} + \frac{1}{\dfrac{1}{j\omega C}} \tag{4.92}$$

Note that impedances in parallel add just like resistors in parallel. We saw from Example 33 that impedances in series add just like resistors in series.

Listing 4.3 is a Matlab program that plots the magnitudes of the phasor currents $\mathbf{I}_R$ and $\mathbf{I}_L$ as a function of frequency from 10 kHz to 1 MHz. Matlab Example 17 runs the program for values of $R = 470\ \Omega$, $C = 3$ nF, and $I_M = 2$ mA. The resulting plot is shown in Fig. 4.20. In Listing 4.3, we used Eq. (4.85) to calculate the phasor voltage at 100 different frequencies between 10 kHz and 1 MHz. Note, from the plot in Fig. 4.20, that the current through the capacitor goes up with frequency as expected. The current through the resistor must therefore go down because the sum of these two

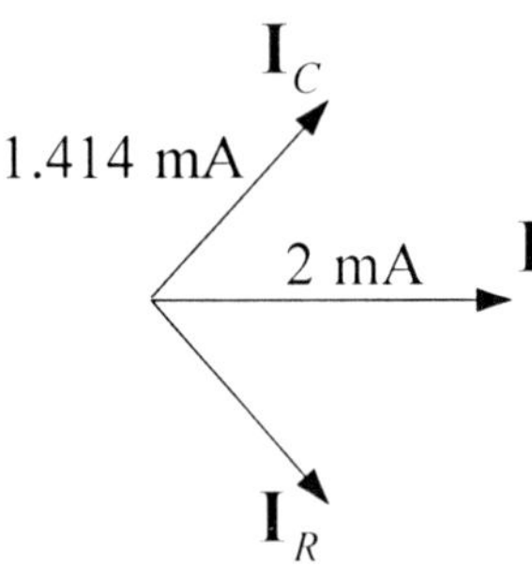

Figure 4.19
The phasor currents when
$I_R = I_C$ in Fig. 4.18

currents must equal the input current of 2 mA. Note again that, just as in Example 16, it is the *phasor* currents that must be added together and these two currents will be 90 degrees out of phase with each other as shown in Fig. 4.19 for the case when the magnitudes of I_R and I_C are equal.

Listing 4.3 RC.m

```
function RC(R,C,I)
% The current I goes through R and C in parallel
% plot magnitude of Ir and Ic as function of frequency
%       from 10kHz to 1MHZ
% R = resistance in ohms
% C = capacitance in farads
% I = current in amps
% RC(R,C,I)
```

Listing 4.3 (cont.) RC.m

```
f = linspace(10000,1000000,100);
w = 2*pi.*f;
V = I./((1/R) + j*w.*C);
Ir = V./R;
Ic = j.*w.*C.*V;

mIr = abs(Ir);
mIc = abs(Ic);
curves = [mIr; mIc];
plot(f,curves)
```

Matlab Example 17

```
>> help RC
  The current I goes through R and C in parallel
  plot magnitude of Ir and Ic as function of frequency
       from 10kHz to 1MHZ
  R = resistance in ohms
  C = capacitance in farads
  I = current in amps
  RC(R,C,I)

>> R = 470;
>> C = 0.000000003;
>> I = 0.002;
>> RC(R,C,I)
```

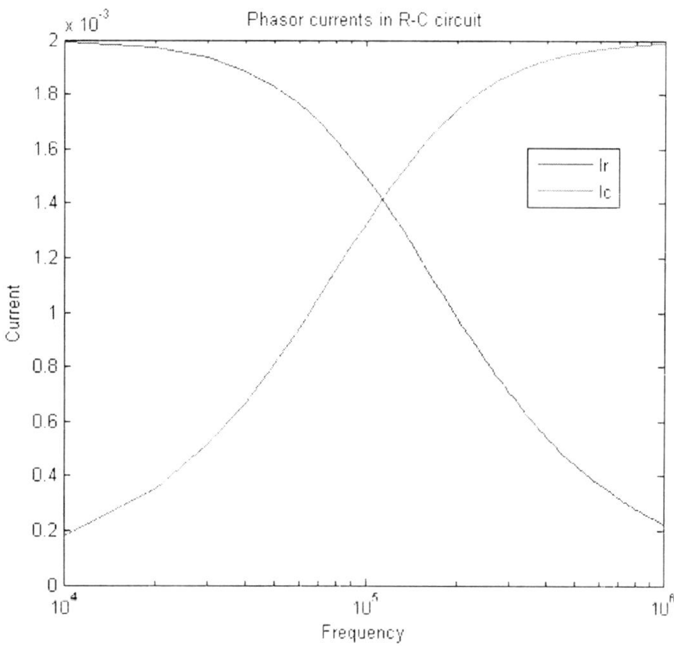

Figure 4.20 Plot resulting from Matlab Example 17

Problems

4.1   Find the rms value of the fullwave rectified sine wave voltage waveform shown below.

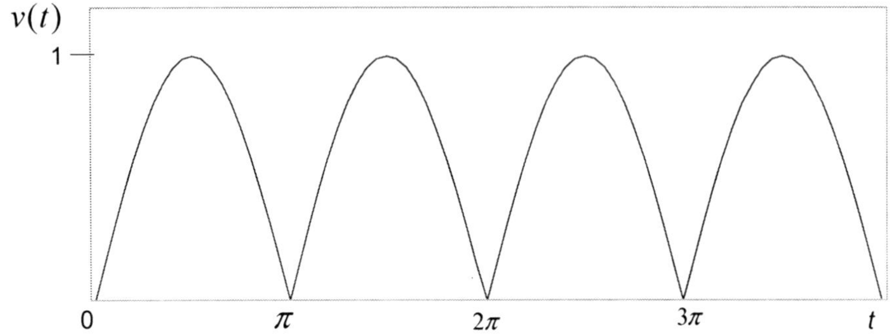

4.2   Find the rms value of the sawtooth voltage waveform shown below.

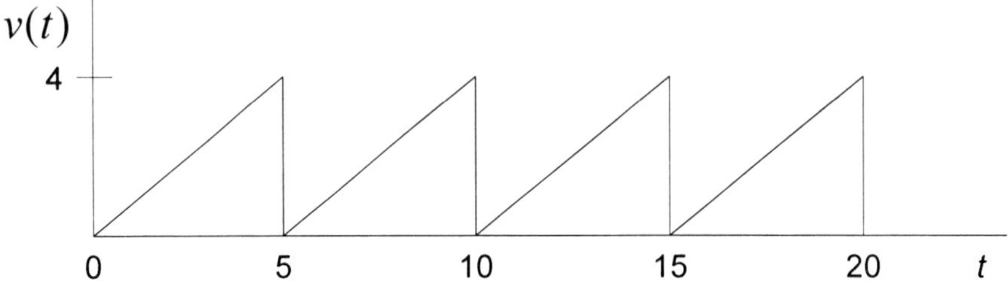

Chapter 5

Basic Circuit Analysis

Using Ohm's law and Kirchhoff's laws, we can solve any DC or steady-state AC circuit problem. For AC circuits, we use the complex impedance instead of simple resistors. Using these basic laws, we can derive various short cuts that will make it easier to solve circuit problems. Three such short cuts will be introduced in this chapter: voltage divider, current division, and source transformation. In Chapter 7, we will show you how to solve any circuit problem by inspection using mesh and nodal analysis. Superposition and Thevenin's and Norton's theorems will be introduced in Chapter 8.

5.1 Voltage Divider

Suppose you have a known voltage $v0$ across two resistors in series as shown in Fig. 5.1. It is easy to determine the voltage across each resistor as a fraction of the voltage across both resistors. The circuit is called a *voltage divider*.

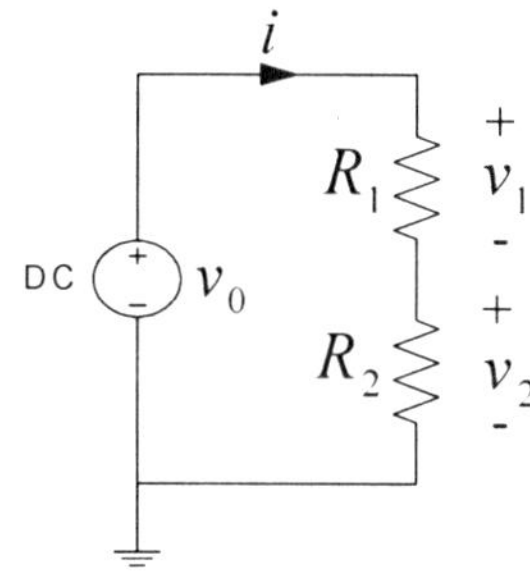

Figure 5.1 A voltage divider

The current i flows through both resistors and from Ohm's law is given by

$$i = \frac{v_0}{R_1 + R_2}$$

(5.1)

The voltage v_2 is then given by

$$v_2 = iR_2 = \frac{v_0}{R_1 + R_2} R_2$$

(5.2)

or

$$v_2 = \frac{R_2}{R_1 + R_2} v_0 \tag{5.3}$$

In a similar way, you can show that

$$v_1 = \frac{R_1}{R_1 + R_2} v_0 \tag{5.4}$$

Thus, the voltage v_0 is divided between the two resistors in proportion to their resistances.

Example 18 – Voltage Divider

In the circuit shown in Fig. 5.2, if $r1 = 2\Omega$, $r2 = 4\Omega$, $r3 = 6\Omega$, and $r4 = 8\Omega$, find the voltage $v1$ and the four currents $i1$, $i2$, $i3$, and $i4$.

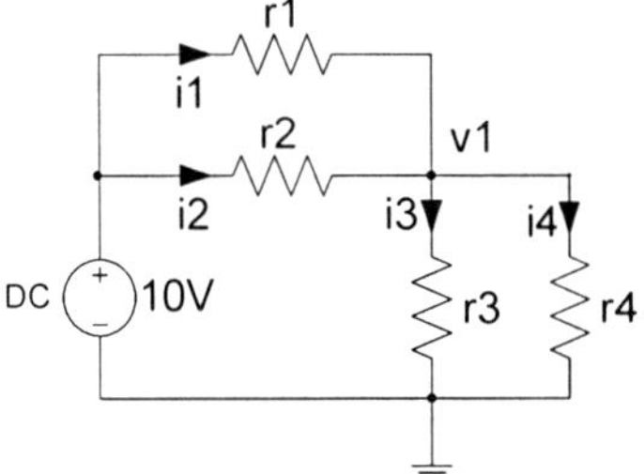

Figure 5.2 Example 18

The resistors $r1$ and $r2$ are in parallel and can be replaced with an equivalent resistance $r5 = \dfrac{r1 \times r2}{r1 + r2}$. Similarly, resistors $r3$ and $r4$ are in parallel and can be replaced with an equivalent resistance $r6 = \dfrac{r3 \times r4}{r3 + r4}$. The resulting circuit shown in Fig. 5.3 is a voltage divider and the voltage $v1 = \dfrac{r6}{r5 + r6} 10V = 7.2V$. Knowing the voltage $v1$ the four currents in the original circuit can be found using Ohm's law as shown in Matlab Example 18. Note that the sum of $i1$ and $i2$ is equal to the sum of $i3$ and $i4$. Why?

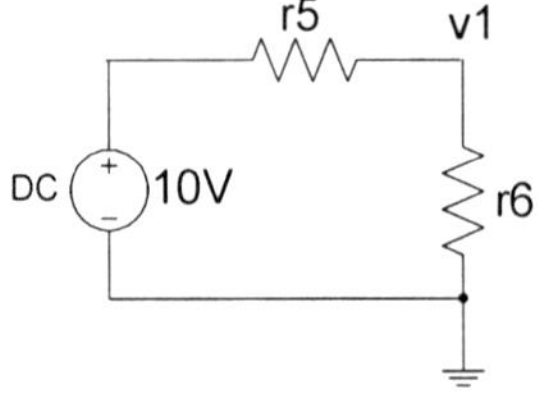

Figure 5.3 Example 18 – step 2

Matlab Example 18

```
>> r1=2;
>> r2=4;
>> r3=6;
>> r4=8;
>> r5 = (r1*r2)/(r1+r2)
r5 =
     1.3333

>> r6 = (r3*r4)/(r3+r4)
r6 =
     3.4286

>> v1 = (r6/(r5+r6))*10
v1 =
     7.2000

>> i1 = (10-v1)/r1
i1 =
     1.4000

>> i2 = (10-v1)/r2
i2 =
     0.7000

>> i3 = v1/r3
i3 =
     1.2000

>> i4 = v1/r4
i4 =
     0.9000
```

Example 19 – R-L Series Circuit

The circuit shown in Fig. 5.4 is the same as that in Example 16. If $R = 1$ kΩ, $L = 1$ mH, and $V_M = 2$ V, find the phasor voltages $\mathbf{V}_R$ and $\mathbf{V}_L$ at a frequency of 100 kHz.

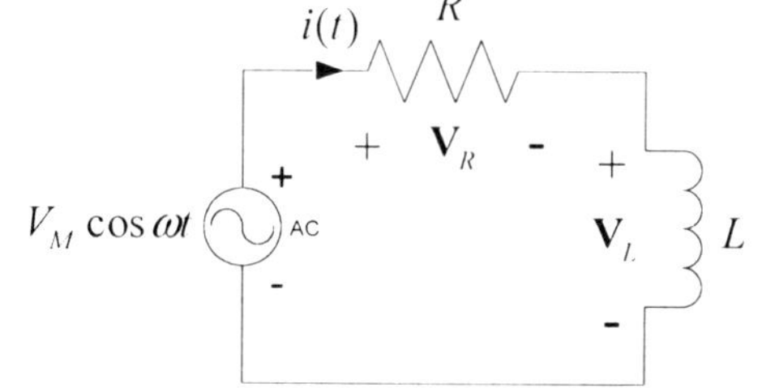

Figure 5.4 A series R-L circuit

The input phasor voltage is given by

$$\mathbf{V} = 2\angle 0^\circ \text{ V} \qquad (5.5)$$

The impedance of the inductor is $j\omega L$. Using the voltage divider equations, we can immediately write

$$\mathbf{V}_L = \mathbf{V}\,\frac{j\omega L}{R + j\omega L} \qquad (5.6)$$

and

$$\mathbf{V}_R = \mathbf{V}\frac{R}{R + j\omega L} \tag{5.7}$$

Matlab Example 19 shows the solution to this problem. The phasor diagram is shown in Fig. 5.5. Note that $\mathbf{V}_L$ leads $\mathbf{V}_R$ (and therefore the current) by 90°.

Matlab Example 19

```
>> R = 1000;
>> L = 10^-3;
>> Vm = 2;
>> f = 100000;
>> w = 2*pi*f;
>> VL = Vm*j*w*L/(R+j*w*L)
VL =
   0.5661 + 0.9010i

>> VLm = abs(VL)
VLm =
    1.0640

>> VLa = angle(VL)*180/pi
VLa =
   57.8581

>> VR = Vm*R/(R+j*w*L)
VR =
   1.4339 - 0.9010i

>> VRm = abs(VR)
VRm =
    1.6935

>> VRa = angle(VR)*180/pi
VRa =
  -32.1419
```

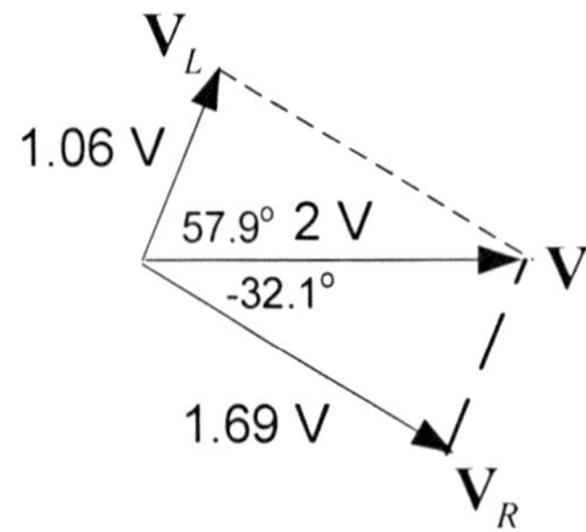

Figure 5.5 Phasor diagram for Example 19

Example 20 – R-C Series Circuit

Consider the R-C series circuit shown in
Fig. 5.6. If $R = 470\ \Omega$, $C = 3$ nF, and V_M
= 2V, find the phasor voltages $\mathbf{V}_R$ and $\mathbf{V}_C$
at a frequency of 100 kHz.

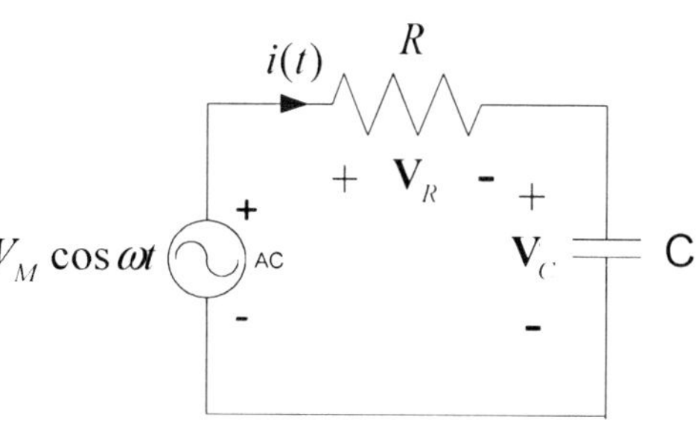

Figure 5.6 A series R-C circuit

Using the impedance of the capacitor, we
can immediately write the voltage divider
equations as

$$\mathbf{V}_C = \mathbf{V}\,\frac{-j/\omega C}{R - j/\omega C} \tag{5.8}$$

and

$$\mathbf{V}_R = \mathbf{V}\,\frac{R}{R - j/\omega C} \tag{5.9}$$

Matlab Example 20 shows the solution to this problem. The phasor diagram is shown in
Fig. 5.7. Note that $\mathbf{V}_R$ (and therefore the current) leads $\mathbf{V}_C$ by 90°.

Matlab Example 20

```
>> R = 470;
>> C = 3*10^-9;
>> Vm = 2;
>> f = 100000;
>> w = 2*pi*f;
>> VC = Vm*(-j/(w*C))/(R-j/(w*C))
VC =
    1.1205 - 0.9927i

>> VCm = abs(VC)
VCm =
     1.4970

>> VCa = angle(VC)*180/pi
VCa =
   -41.5387

>> VR = Vm*R/(R-j/(w*C))
VR =
    0.8795 + 0.9927i

>> VRm = abs(VR)
VRm =
     1.3263

>> VRa = angle(VR)*180/pi
VRa =
    48.4613
```

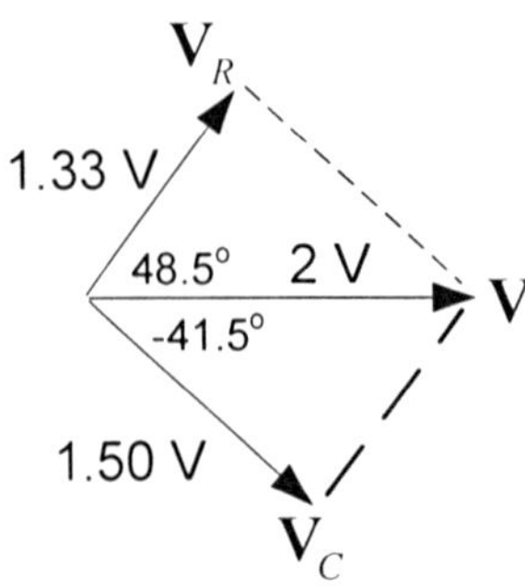

Figure 5.7 Phasor diagram for Example 20

Example 21 – RLC Circuit

For the circuit shown in Fig. 5.8, plot the magnitude of the output voltage V_R as a function of frequency from 10 kHz to 1 MHz. Let $R = 600\ \Omega$, $C = 2.653$ nF, $L = 9.5$ mH, and $V_M = 1$ V.

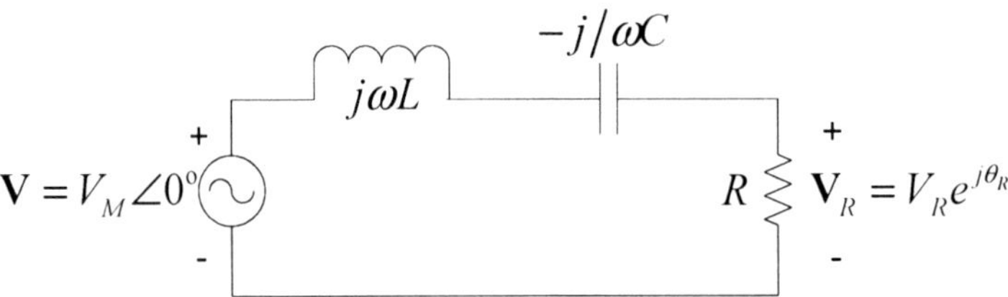

Figure 5.8 Example of an RLC circuit

First let's see if you can picture what the magnitude of the output voltage V_R will look like as a function of frequency. At zero frequency, no current flows through the capacitor so the voltage across the resistor must be zero. At very high frequencies, the voltage across the inductor will grow meaning that the voltage across the resistor will decrease. We would expect the maximum voltage across the resistor to occur at some intermediate frequency.

Thinking in terms of impedances, the circuit in Fig. 5.8 is just a voltage divider. Therefore, we can immediately write the phasor voltage across the resistor as

$$\mathbf{V}_R = \frac{R}{R + j\omega L - j/\omega C}\mathbf{V} = \frac{R}{R + j(\omega L - 1/\omega C)}\mathbf{V} \qquad (5.10)$$

Note that the output voltage $\mathbf{V}_R$ will be a maximum and equal to the input voltage $\mathbf{V}$ when $\omega L = 1/\omega C$. This will occur at the angular frequency

$$\omega_0 = \frac{1}{\sqrt{LC}}\ \text{rad/s} \qquad (5.11)$$

or, since $\omega_0 = 2\pi f_0$,

$$f_0 = \frac{1}{2\pi\sqrt{LC}}\,\text{Hz} \tag{5.12}$$

The frequency ω_0 (or f_0) is called the *resonant* frequency and occurs when the inductive reactance exactly cancels out the capacitive reactance leaving the circuit purely resistive.

Listing 5.1 is a Matlab program that plots the magnitude of the phasor voltage $\mathbf{V}_R$ as a function of frequency from 10 kHz to 1 MHz. Matlab Example 21 runs the program for values of $R = 600\ \Omega$, $L = 9.5$ mH, $C = 2.653$ nF, and $V_M = 1$ V. The resulting plot is shown in Fig. 5.9. In Listing 5.1, we used Eq. (5.10) to calculate the phasor voltage at 100 different frequencies between 10 kHz and 1 MHz. Note from the plot in Fig. 5.9 that the maximum output occurs at the resonant frequency given by Eq. (5.12).

Listing 5.1 RLC.m

```
function RLC(R,L,C,V)
% The voltage V is across L, C, and R in series
% plot magnitude of Vr as function of frequency
%      from 10kHz to 1MHZ
% R = resistance in ohms
% L = inductance in henrys
% C = capacitance in farads
% V = voltage in volts
% RL(R,L,C,V)
f = linspace(10000,1000000,100);
w = 2*pi.*f;
Vr = R.*V./(R + j.*(w.*L-1./(w.*C)));
mVr = abs(Vr);
plot(f,mVr)
```

Matlab Example 21

```
>> help RLC
  The voltage V is across L, C, and R in series
  plot magnitude of Vr as function of frequency
        from 10kHz to 1MHZ
  R = resistance in ohms
  L = inductance in henrys
  C = capacitance in farads
  V = voltage in volts
  RL(R,L,C,V)

>> R = 600;
>> L = 0.0095;
>> C = 0.2653*10^-9;
>> V = 1;
>> RLC(R,L,C,V)
```

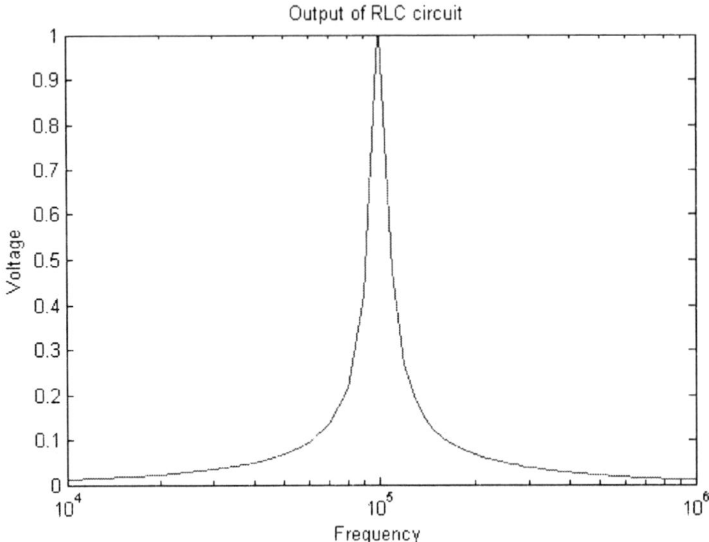

Figure 5.9 Plot resulting from Matlab Example 21

Circuits of this type can be used as *filters* to filter out high and low frequencies and let through only a band of frequencies. Circuits containing more inductors and capacitors can produce filters with narrower pass bands and sharper cutoff frequencies.

5.2 Wheatstone Bridge

The circuit shown in Fig. 5.10 is called a Wheatstone bridge. It can be used to measure an unknown resistance R_3 (the arrow indicates that the resistance R_3 is variable).

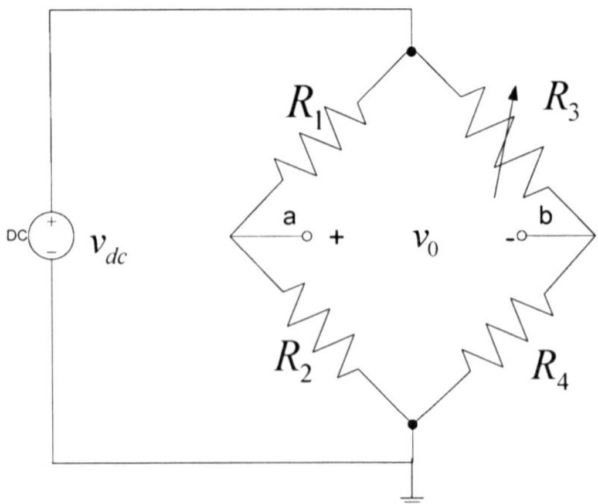

Figure 5.10 Wheatstone bridge

Note that each side of the bridge forms a voltage divider. Thus, the voltage at point a is given by

$$v_a = v_{dc}\left(\frac{R_2}{R_1 + R_2}\right) \tag{5.13}$$

and the voltage at point b is given by

$$v_b = v_{dc}\left(\frac{R_4}{R_3 + R_4}\right) \tag{5.14}$$

Note that, if $R_1 = R_2$ and $R_3 = R_4$, then $v_0 = v_a - v_b = 0$. In general, we can write

$$v_0 = v_a - v_b$$

$$v_0 = v_{dc}\left(\frac{R_2}{R_1 + R_2} - \frac{R_4}{R_3 + R_4}\right) \tag{5.15}$$

$$\frac{v_0}{v_{dc}} = \frac{R_2}{R_1 + R_2} - \frac{R_4}{R_3 + R_4} \tag{5.16}$$

Multiplying each side of Eq. (5.11) by $(R_3 + R_4)$ and rearranging, we get

$$(R_3 + R_4)\left(\frac{v_0}{v_{dc}} - \frac{R_2}{R_1 + R_2}\right) = -R_4$$

and solving for R_3 gives

$$R_3 = R_4\left[\frac{1}{\left(\dfrac{R_2}{R_1 + R_2} - \dfrac{v_0}{v_{dc}}\right)} - 1\right] \tag{5.17}$$

In many applications, the resistance R_3 is a sensor such as a strain gauge in which the resistance changes due to a change in strain. Recall that, if $R_1 = R_2$ and $R_3 = R_4$, then the bridge is balanced and $v_0 = v_a - v_b = 0$. Small changes in R_3 will cause v_0 to change.

Example 22 – Wheatstone Bridge

To see how v_0 changes when R_3 changes in the Wheatstone bridge in Fig. 5.10, let $R_1 = R_2 = R_4 = 100\ \Omega$, $R_3 = 100\ \Omega + \Delta R$, and $v_{dc} = 6$ V. Then Eq. (5.15) for v_0 becomes

$$v_0 = 6\left(\frac{100}{200} - \frac{100}{200 + \Delta R}\right) = 3\left(1 - \frac{1}{1 + \dfrac{\Delta R}{200}}\right) \tag{5.18}$$

Note that, if $\Delta R = 0$, then $v_0 = 0$ as expected. If $\Delta R = \infty$ (i.e. an open circuit), then $v_0 = 3$ V because $v_a = 3$ V and $v_b = 0$ V (since no current flows through R_4).

We would like to use Matlab to plot v_0 vs. ΔR. Matlab Example 22a will do this for values of ΔR from 0 to 10 ohms and the resulting plot is shown in Fig. 5.11.

Matlab Example 22a

```
>> delr = 0:10
delr =
     0     1     2     3     4     5     6     7     8     9    10

>> v0 = 3 * (1 - (1 ./ (1 + delr / 200)))
v0 =
  Columns 1 through 7
        0     0.0149     0.0297     0.0443     0.0588     0.0732     0.0874
  Columns 8 through 11
    0.1014     0.1154     0.1292     0.1429

>> plot(delr,v0)
```

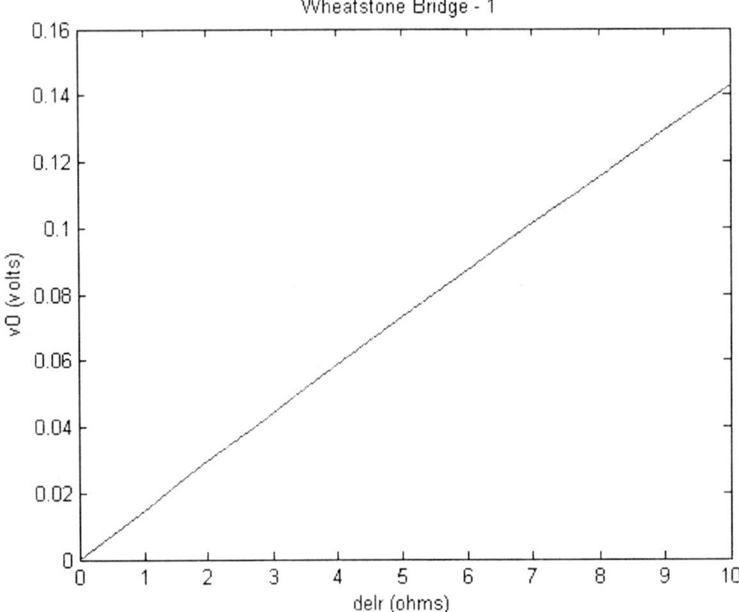

Figure 5.11 v_0 vs. *delr* is linear for values of *delr* less than about 10

In Matlab Example 22a, we wrote

$$\text{delr = 0:10} \tag{5.19}$$

This statement will generate a 1 x 11 row matrix called *delr* that contains the integers 0 through 10 as shown in Matlab Example 22a. To compute a value of v_0 for each value of *delr* using Eq. (5.18) we write the Matlab statement

$$\text{v0 = 3 * (1 - (1 ./ (1 + delr / 200)))} \tag{5.20}$$

We must be careful in writing this equation in Matlab. If we divide the row matrix *delr* by 200, each element of *delr* will be divided by 200, which is what we want. If we then add 1 to this result, each element will be incremented by 1. However, we now want form a new row matrix in which each element is equal to 1 divided by its current value. Recall (see Appendix D) that multiplying matrices requires that the matrices have compatible numbers of rows and columns, and dividing by a matrix really means taking the inverse of the matrix and then multiplying matrices. However, in Eq. (5.20), we really don't want to divide 1 by a row matrix, we want to form a new row matrix in which each element is equal to 1 divided by its current value. Recall that the *dot-division operator ./* will do this. Therefore, the Matlab equation given by Eq. (5.20) will create a row matrix *v0* that contains 11 elements corresponding to each of the 11 elements in *delr*. We can plot v_0 vs. *delr* by using the Matlab statement

$$\texttt{plot(delr,v0)} \qquad (5.21)$$

as shown in Matlab Example 22a. The resulting plot is shown in Fig. 5.11. Note that v_0 is a fairly linear function of *delr* for values of *delr* less than 10 (i.e. about a 10% change in resistance).

Suppose we want *delr* to vary from 0 to 100 in steps of 10. The Matlab statement

$$\texttt{delr = 0:10:100} \qquad (5.22)$$

will do this as shown in Matlab Example 22b. Note from the resulting plot in Fig. 5.12 that $v_0 = 1$ V when *delr* = 100 and the curve is becoming nonlinear. Finally in Matlab Example 22c we let *delr* vary from 0 to 1000 in steps of 50. The resulting plot in Fig. 5.13 shows that the curve is becoming quite nonlinear and the value of v_0 is approaching its limiting value of 3 volts.

Matlab Example 22b

```
>> delr = 0:10:100
delr =
     0    10    20    30    40    50    60    70    80    90   100

>> v0 = 3 * (1 - (1 ./ (1 + delr / 200)))
v0 =
  Columns 1 through 7
        0    0.1429    0.2727    0.3913    0.5000    0.6000    0.6923
  Columns 8 through 11
    0.7778    0.8571    0.9310    1.0000

>> plot(delr,v0)
```

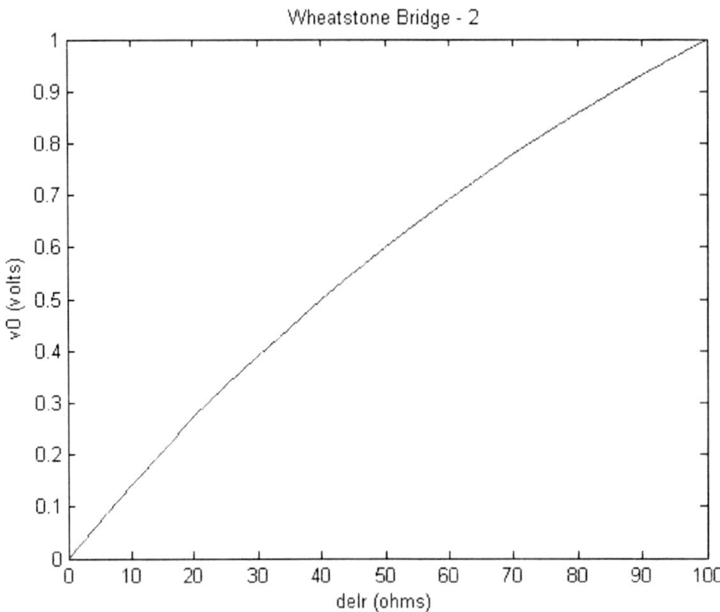

Figure 5.12 *v0* vs. *delr* is becoming nonlinear for larger values of *delr*

Matlab Example 22c

```
>> delr = 0:50:1000;

>> v0 = 3 * (1 - (1 ./ (1 + delr / 200)));

>> plot(delr,v0)
```

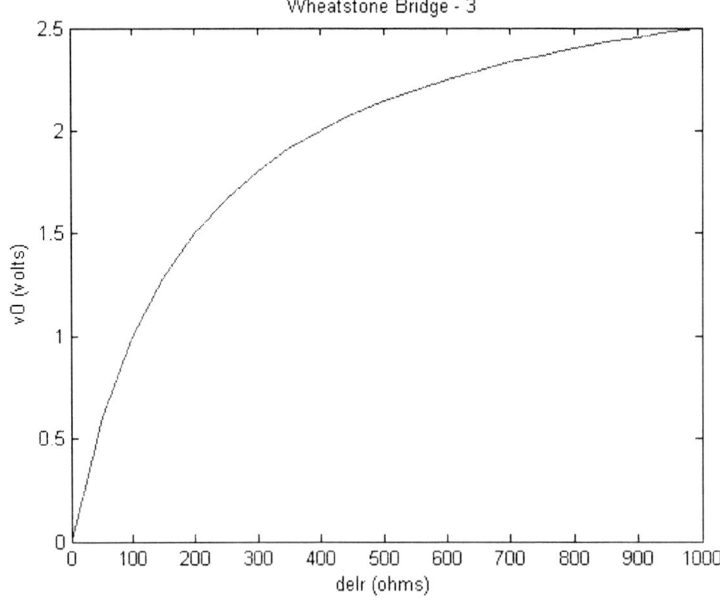

Figure 5.13 For very large values of *delr v0* vs. *delr* is very nonlinear and *v0* approaches 3V

5.3 Current Division

Suppose you have a known current i that is split between two resistors in parallel as shown in Fig. 5.14. It is easy to determine the current through each resistor as a fraction of the total current i. This is called *current division*.

The same voltage v appears across both resistors. This voltage is equal to the current i times the equivalent parallel resistance, R_p. That is,

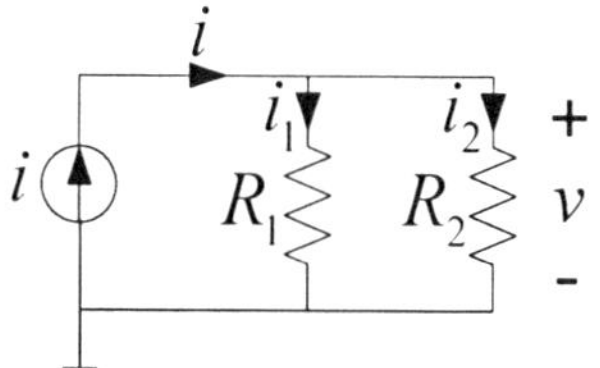

$$v = R_p i = \frac{R_1 R_2}{R_1 + R_2} i \qquad (5.23)$$

Figure 5.14 Current division

Using this result and Ohm's law, we can write equations for the currents i_1 and i_2 through each resistor. Thus,

$$i_1 = \frac{v}{R_1} = \frac{R_2}{R_1 + R_2} i \qquad\qquad i_2 = \frac{v}{R_2} = \frac{R_1}{R_1 + R_2} i \qquad (5.24)$$

We therefore see that the current divides in inverse proportion to the resistances. Thus, to find the current through R_1, you would take the other resistance, R_2, and divide it by the sum of the resistances to find the fraction of the total current i that goes through R_1. Similarly, to find the current i_2, you would divide R_1 by the sum of the resistances.

If you have N resistors in parallel, we saw in Eq. (2.16) that the equivalent parallel resistance would be given by

$$\frac{1}{R_P} = \frac{1}{R_1} + \frac{1}{R_2} + \dots + \frac{1}{R_N} \qquad (5.25)$$

The voltage across these N resistors would be

$$v = R_p i \qquad (5.26)$$

The current in the jth branch would then be given by

$$i_j = \frac{v}{R_j} = \frac{R_P}{R_j} i \qquad (5.27)$$

Example 23 – Current Division

In the circuit shown in Fig. 5.15, if $r1 = 2\Omega$, $r2 = 4\Omega$, $r3 = 6\Omega$, and $r4 = 8\Omega$, find the four currents $i1$, $i2$, $i3$, and $i4$, and the two voltages $v0$ and $v1$.

The 6A current source is split between the two parallel resistors $r1$ and $r2$. Using current division, $i1 = \dfrac{r2}{r1+r2}6A = 4A$ and $i2 = \dfrac{r1}{r1+r2}6A = 2A$. The 6A current source is also split between the two parallel resistors $r3$ and $r4$. Using current division, $i3 = \dfrac{r4}{r3+r4}6A = 3.43A$ and $i4 = \dfrac{r3}{r3+r4}6A = 2.57A$. The voltages $v1$ and $v0$ can then be found using Ohm's law as shown in Matlab Example 23. Using either $i3$ and $r3$ or $i4$ and $r4$ we find that $v1 = 20.57$ V. Calculating the voltage across either $r1$ or $r2$ we find that $v0 = 28.57$ V.

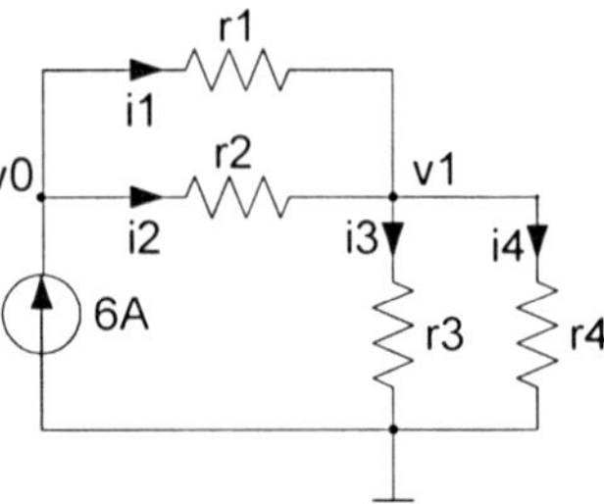

Figure 5.15 Example 23

Matlab Example 23

```
>> r1=2;
>> r2=4;
>> r3=6;
>> r4=8;
>> i1 = (r2/(r1+r2))*6
i1 =

     4

>> i2 = (r1/(r1+r2))*6
i2 =

     2

>> i3 = (r4/(r3+r4))*6
i3 =

     3.4286

>> i4 = (r3/(r3+r4))*6
i4 =

     2.5714

>> v1 = i3*r3
v1 =

    20.5714

>> v1 = i4*r4
v1 =

    20.5714

>> v0 = v1 + i1*r1
v0 =

    28.5714

>> v0 = v1 + i2*r2
v0 =

    28.5714
```

Example 24 – R-C Parallel Circuit

The circuit shown in Fig. 5.16 is the same circuit as in Example 17. Use current division to find the phasor currents $\mathbf{I}_R$ and $\mathbf{I}_C$, and then find the voltage $v(t)$. If $R = 470 \ \Omega$, $C = 3$ nF, and $I_M = 2$ mA, plot the phasor diagram at a frequency of 100 kHz.

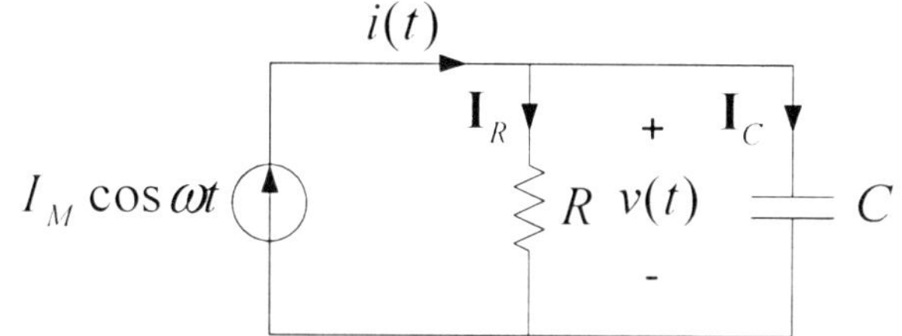

Figure 5.16 A parallel R-C circuit

The impedance of the capacitor is

$$Z_C = \frac{-j}{\omega C} = \frac{-j}{2\pi fC} = \frac{-j}{2\pi 10^5 \times 3 \times 10^{-9}} = -j530.5 \qquad (5.28)$$

The phasor current source is

$$\mathbf{I} = 2 \times 10^{-3} \angle 0^\circ \ \text{A} \qquad (5.29)$$

Using current division, we can then write

$$\mathbf{I}_R = \mathbf{I}\frac{-j/\omega C}{R - j/\omega C} = 2 \times 10^{-3} \angle 0^\circ \frac{-j530.5}{470 - j530.5}$$

$$\mathbf{I}_R = 1.121 - j0.993 = 1.497 \angle -41.5^\circ \ \text{mA} \qquad (5.30)$$

and

$$\mathbf{I}_C = \mathbf{I}\frac{R}{R - j/\omega C} = 2 \times 10^{-3} \angle 0^\circ \frac{470}{470 - j530.5}$$

$$\mathbf{I}_C = 0.880 + j0.993 = 1.326 \angle 48.5^\circ \ \text{mA} \qquad (5.31)$$

The phasor voltage, $\mathbf{V}$, is given by

$$\mathbf{V} = R\mathbf{I}_R = 470 \times 1.497 \angle -41.5^\circ = 703.6 \angle -41.5^\circ \ \text{mV} \qquad (5.32)$$

The corresponding voltage $v(t)$ is given by

$$v(t) = 703.6 \cos(2\pi 10^5 t - 41.5^\circ) \ \text{mV}$$

The resulting phasor diagram is shown in Fig. 5.17. Note that the current through the capacitor leads the voltage across the capacitor by 90°.

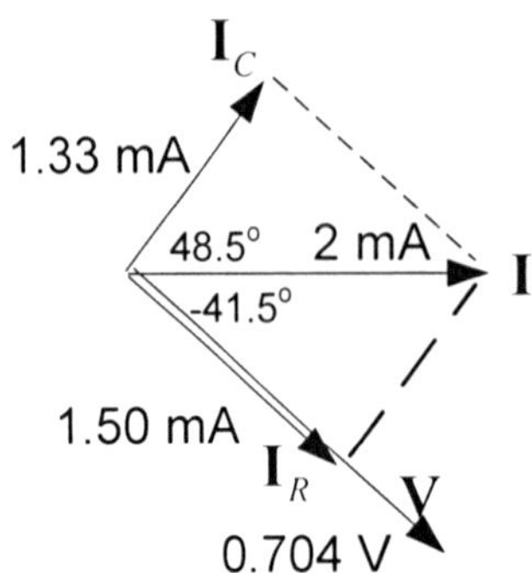

Figure 5.17 Phasor diagram for Example 24

Example 25 – R-L Parallel Circuit

Consider the R-C series circuit shown in Fig. 5.6. Use current division to find the phasor currents $\mathbf{I}_R$ and $\mathbf{I}_L$, and then find the voltage $v(t)$. If $R = 470\ \Omega$, $L = 1$ mH, and $I_M = 2$ mA, plot the phasor diagram at a frequency of 100 kHz.

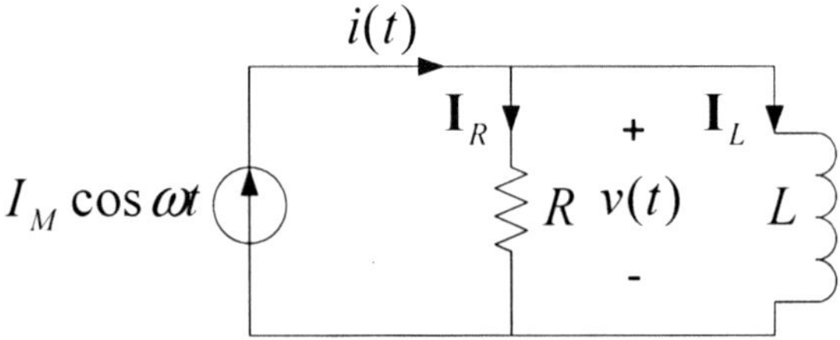

Figure 5.18 Circuit for Example 25

The impedance of the inductor is

$$Z_L = j\omega L = j2\pi fL = j2\pi 10^5 \times 1 \times 10^{-3} = j628.3 \tag{5.33}$$

The phasor current source is

$$\mathbf{I} = 2 \times 10^{-3} \angle 0^\circ \tag{5.34}$$

Using current division, we can then write

$$\mathbf{I}_R = \mathbf{I}\frac{j\omega L}{R + j\omega L} = 2 \times 10^{-3} \angle 0^\circ \frac{j628.3}{470 + j628.3}$$

$$\mathbf{I}_R = 1.282 + j0.959 = 1.601\angle 36.8^\circ \text{ mA} \tag{5.35}$$

and

$$\mathbf{I}_L = \mathbf{I}\frac{R}{R + j\omega L} = 2 \times 10^{-3} \angle 0^\circ \frac{470}{470 + j628.3}$$

$$\mathbf{I}_L = 0.718 + j0.959 = 1.198\angle -53.2^\circ \text{ mA} \tag{5.36}$$

The phasor voltage, **V**, is given by

$$\mathbf{V} = R\mathbf{I}_{R} = 470 \times 1.601\angle 36.8^{\circ} = 752.5\angle 36.8^{\circ} \text{ mV} \qquad (5.37)$$

The corresponding voltage $v(t)$ is given by

$$v(t) = 752.5\cos(2\pi 10^{5} t + 36.8^{\circ}) \text{ mV}$$

The resulting phasor diagram is shown in Fig. 5.19. Note that the voltage across the inductor leads the current through the inductor by 90°.

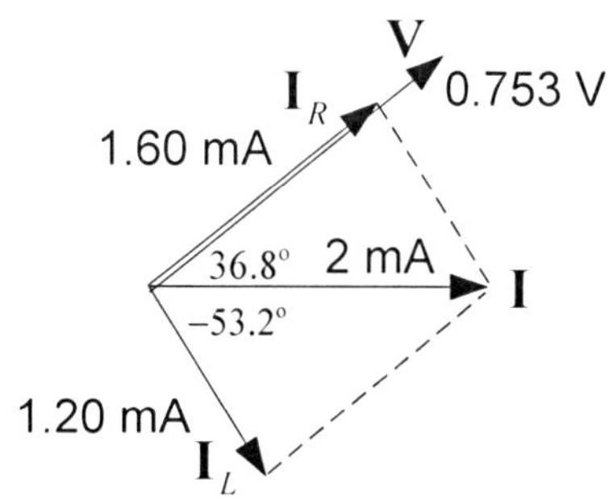

Figure 5.19 Phasor diagram for Example 25

5.4 Source Transformation

We can always replace a voltage source in series with a resistor by a current source in parallel with the same resistor, and vice-versa, as shown in Fig. 5.20. This is called *source transformation*, or *source exchange*. This will sometimes simplify a circuit and make it easier to solve for currents and voltages.

To prove that this works we will show that the voltage across and the current through any load connected to the terminals in the two circuits in Fig. 5.20 will be the same.

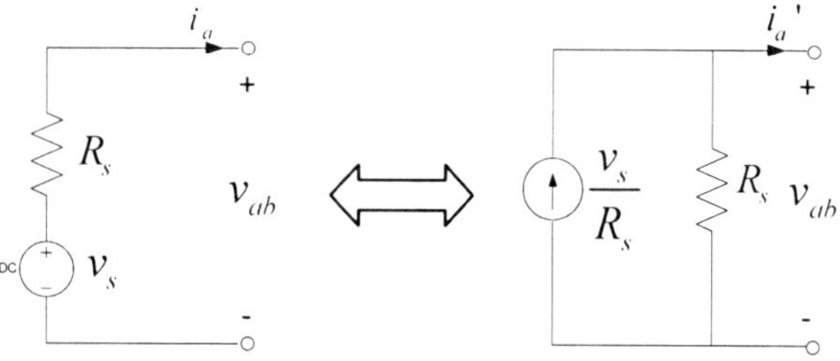

Figure 5.20 Source exchange

In Fig. 5.21, we have connected a load resistor R_L to both circuits in Fig. 5.20. The circuit in Fig. 5.21a is a voltage divider and the voltage v_L across R_L is

$$v_{L} = \frac{R_{L}}{\left(R_{s} + R_{L}\right)} v_{s} \qquad (5.38)$$

The current i_a in Fig. 5.24a will be

$$i_a = \frac{v_s}{\left(R_s + R_L\right)} \qquad (5.39)$$

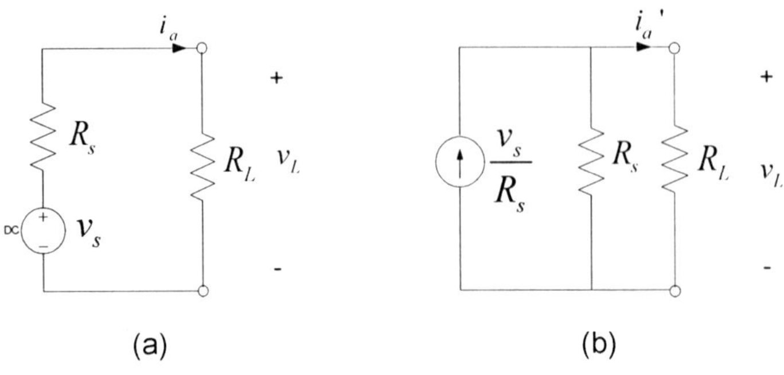

(a)   (b)

Figure 5.21 Source exchange proof

If we use current division to find the current i_a' in Fig. 5.21b, we see that

$$i_a' = \frac{R_s}{\left(R_s + R_L\right)}\frac{v_s}{R_s} = \frac{v_s}{\left(R_s + R_L\right)} \qquad (5.40)$$

which is the same as the current i_a in Eq. (5.39) for the circuit in Fig. 5.21a. The voltage v_L in Fig. 5.21b is given by

$$v_L = i_a'R_L = \frac{R_L}{\left(R_s + R_L\right)}v_s \qquad (5.41)$$

which is the same as the voltage v_L in Eq. (5.38) for the circuit in Fig. 5.21a.

Thus, from the point of view of the load R_L, the two circuits in Fig. 5.21 are equivalent. We say that as seen by the load the two circuits are equivalent.

Example 26 – DC Source Transformation

Find the voltage $v1$ and the currents $i1$ and $i2$ in the circuit shown in Fig. 5.22.

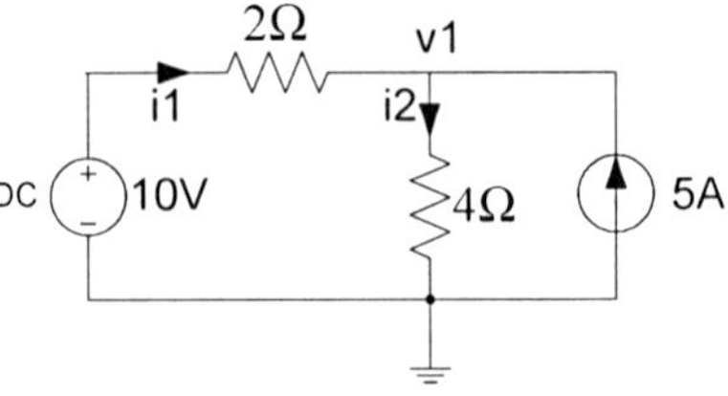

Figure 5.22 Example 26

Using source transformation, we can replace the 5A current source in parallel with the 4Ω resistor with a 20V voltage source in series with a 4Ω resistor as shown in Fig. 5.23. From this circuit, we can find the current $i1$ using Ohm's law.

$$i1 = \left(10V - 20V\right)/6\Omega$$

We can now go back to the circuit in Fig. 5.22 and use source transformation to replace the 10V voltage source in series with the 2Ω resistor with a 5A current source in parallel with a 2Ω resistor as shown in Fig. 5.24.   In this circuit, we can combine the two current sources (by using Kirchhoff's current law from Section 3.2) as shown in Fig. 5.25. Now, in Fig. 5.25, we can use current division to find $i2$.

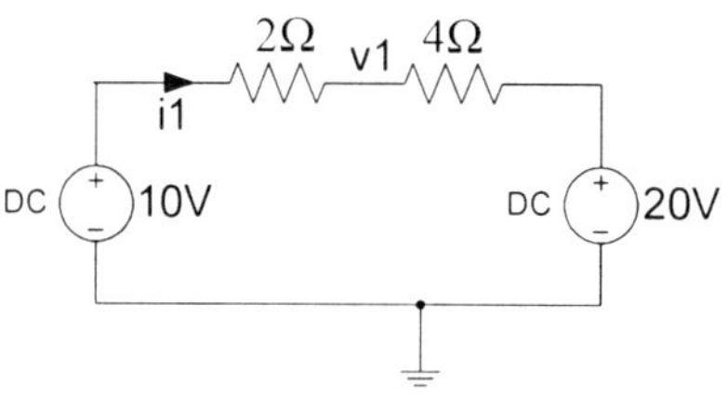

Figure 5.23 Example 26 – step 1

$$i2 = \frac{2\Omega}{2\Omega + 4\Omega} 10A$$

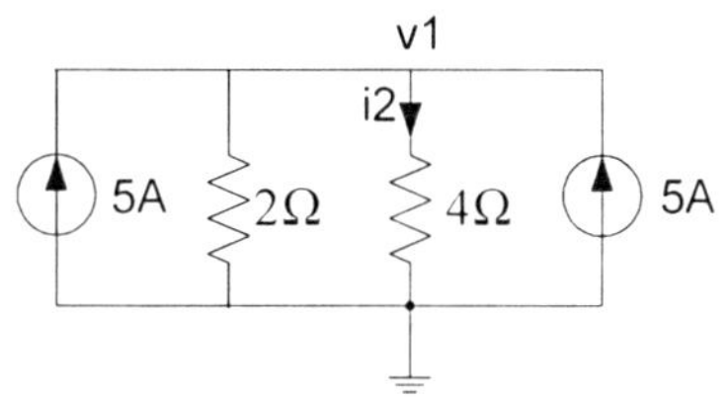

Figure 5.24 Example 26 – step 2

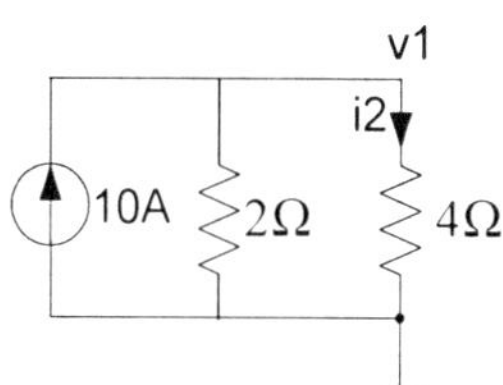

Figure 5.25 Example 26 – step 3

Finally, in the original circuit, the voltage $v1$ is given by $v1 = i2 \times 4\Omega$. These solutions are shown in Matlab Example 26

Matlab Example 26

```
>> i1 = (10-20)/6
i1 =
    -1.6667

>> v1 = 10 - i1*2
v1 =
    13.3333

>> i2 =(2/(2+4))*10
i2 =
     3.3333

>> v1 = i2*4
v1 =
    13.3333
```

Example 27 – Source Transformation

The circuit shown in Fig. 5.26 is the same circuit that we solved in Example 8 using KCL. Let's solve this same circuit using source transformation. Applying source transformation to both of the voltage sources in Fig. 5.26, we obtain the equivalent circuit shown in Fig. 5.27. We can then combine the three parallel resistors to an equivalent parallel resistor, add the two current sources and use Ohm's law to find the voltage $v1$.

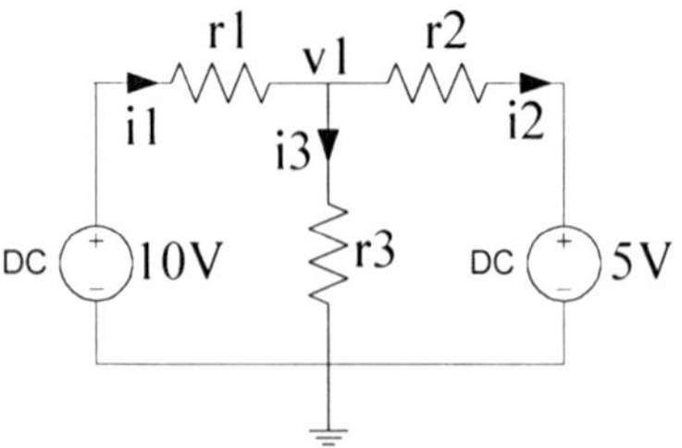

Figure 5.26 Example 27

The Matlab solution to do this is shown in Matlab Example 27 where we converted the resistances to conductances to make it easier to find the parallel equivalent conductance. Note that the voltage $v1$ is the same value found in Example 8. The three currents can be found using Ohm's law as was done in Example 8.

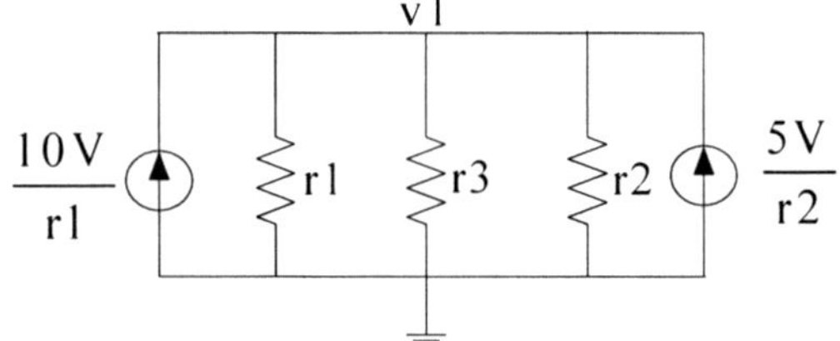

Figure 5.27 Example 27

Matlab Example 27

```
>> r1 = 6;
>> r2 = 3;
>> r3 = 4;
>> g1 = 1/r1;
>> g2 = 1/r2;
>> g3 = 1/r3;
>> gp = g1+g2+g3
gp =
    0.7500

>> i01 = 10/r1
i01 =
    1.6667

>> i02 = 5/r2
i02 =
    1.6667

>> i = i01 + i02
i =
    3.3333

>> v1 = i/gp
v1 =
    4.4444
```

Example 28 – AC Source Transformation

Use source transformation to find the phasor current $\mathbf{I}_0$ in the AC circuit shown in Fig. 5.28. Applying source transformation to the voltage souce in series with the impedance 1-j1 produces the circuit shown in Fig. 5.29. Combining the two current sources in Fig. 5.29 produces the circuit shown in Fig. 5.30 where the total current source is

$$\mathbf{I}_T = 2 + \frac{6}{1 - j1} \text{ A} \qquad (5.42)$$

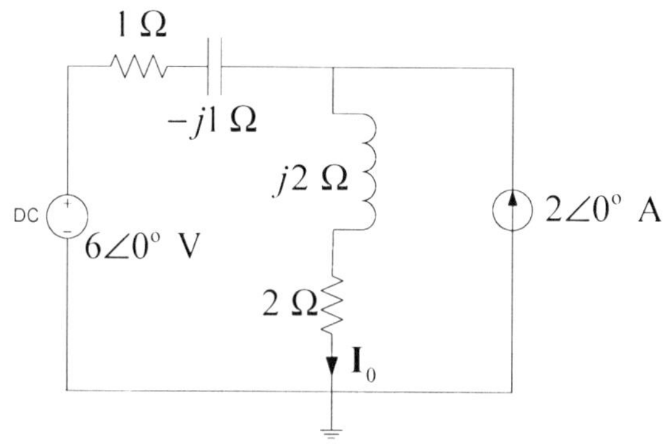

Figure 5.28 Example 28

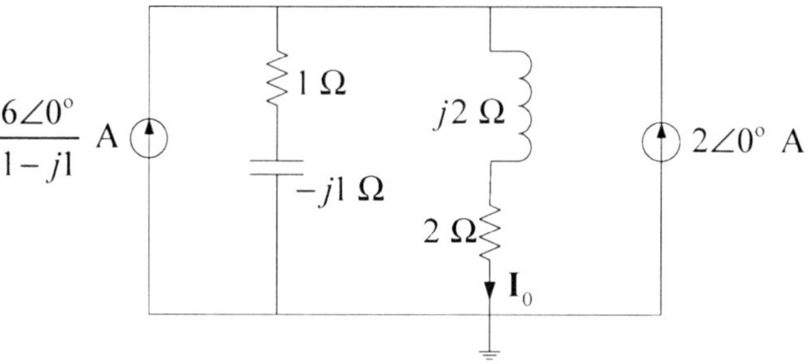

Figure 5.29 Example 28 – Step 1

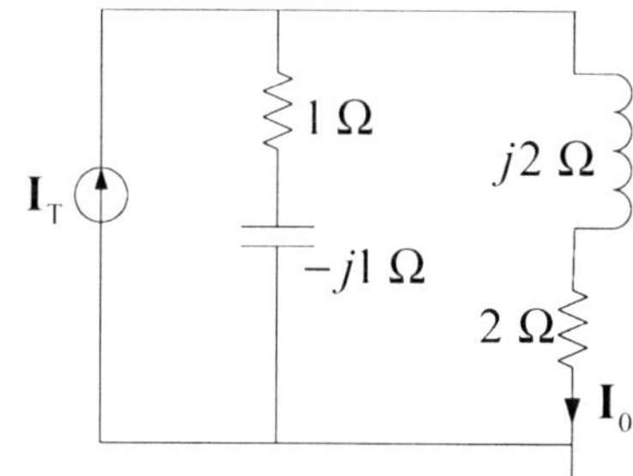

Figure 5.30 Example 28 – Step 2

Using current division, we can write $\mathbf{I}_0$ in Fig. 5.30 as

$$\mathbf{I}_0 = \frac{1 - j1}{1 - j1 + 2 + j2}\mathbf{I}_T = 2.608\angle -32.47° \text{ A}$$

where the result is found in Matlab Example 28.

Matlab Example 28

```
>> It = 2 + 6/(1-1j);

>> I0 = It*(1-1j)/(3+1j);

>> magI0 = abs(I0)
magI0 =
     2.6077

>> angI0 = angle(I0)*180/pi
angI0 =
   -32.4712
```

Problems

5.1   In the following circuit, find the four currents $i1$, $i2$, $i3$, and $i4$, and the two voltages $v0$ and $v1$.

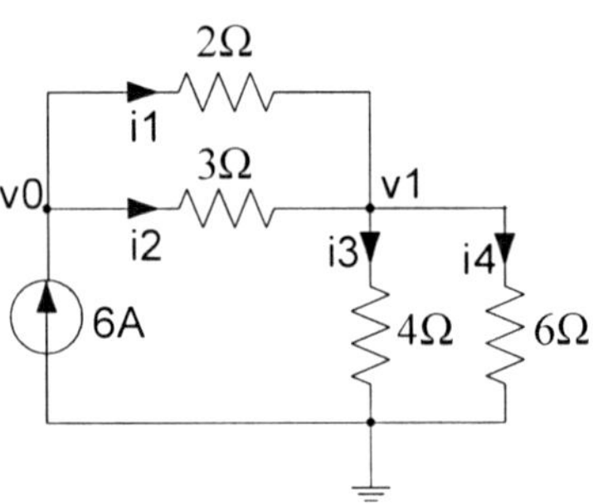

5.2   Using source transformation, find the voltage $v1$ and the currents $i1$ and $i2$ in the following circuit.

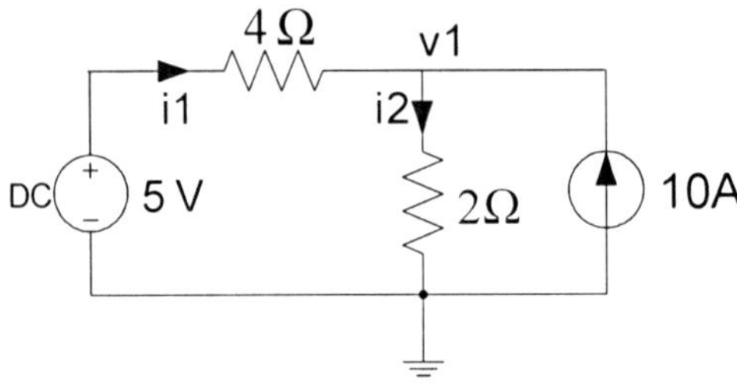

5.3   Use voltage division to find v_0 in the following circuit.

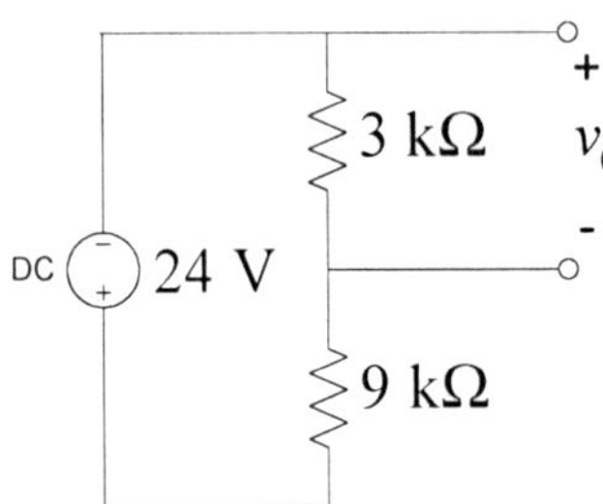

5.4   Use current division to find i_1 in the following circuit.

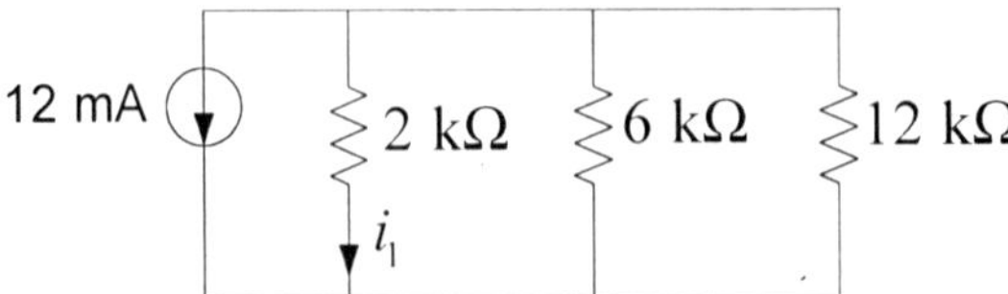

Chapter 6

Operational Amplifiers (Op Amps)

Up to now, all of our circuits have contained only passive elements (resistors) that dissipate power supplied by voltage or current sources. Sometimes we need to amplify signals. Amplifiers require some kind of active device such as transistors. A transistor is able to amplify signals by having a small current (or voltage) control a larger current (or voltage) with the power coming from some external power supply.

A particularly useful kind of amplifier is a solid-state device called an *operational amplifier*, or *op amp*. The original operational amplifiers were made out of vacuum tubes and used as the basic element in analog computers. Analog computers were designed to solve differential equations and the operational amplifiers were high-gain dc amplifiers that could perform certain *operations* (hence the name) such as addition, integration, and multiplication by a constant – just the operations needed to solve differential equations. Analog computers have now been largely replaced with high-speed digital computers but the operational amplifier has developed into one of the most popular electronic devices for designing a wide range of electronic circuits.

One of the first solid state op amps, which sold for about \$300, was the μA702 produced by Fairchild Semiconductor in 1963. Typical op amps used in the lab today cost between \$1 and \$2. In this chapter, we will define an ideal op amp and show how you can add external resistors to an ideal op amp to produce useful amplifier circuits.

6.1 Ideal Op Amp

The symbol for an ideal op amp is shown in Fig. 6.1. An ideal op amp has the following two properties:

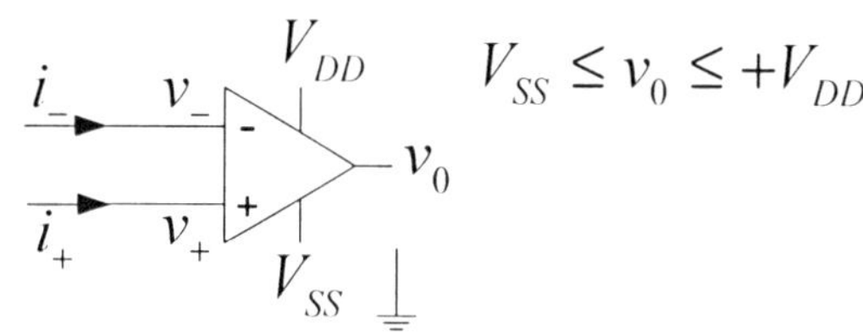

$$V_{SS} \leq v_0 \leq +V_{DD}$$

1) $v_0 = A_v \left(v_+ - v_- \right)$

The open-loop gain, A_v, is very large, approaching infinity.

Figure 6.1 An ideal op amp

2) $i_+ = i_- = 0$

The current into the inputs are zero.

In Fig. 6.1, the supply voltage V_{DD} is typically +2.5V to +15V and the voltage V_{SS} is often a negative voltage equal to $-V_{DD}$. Many newer op amps can operate with a single-sided power supply in which $V_{SS} = 0\text{V}$. Note that the output voltage v_0 of a real op amp must be in the range $V_{ss} \leq v_0 \leq +V_{DD}$. Since the values of V_{ss} and V_{DD} do not affect our analysis of op amp circuits, we adopt standard practice and do not show these values in what follows.

An op amp circuit will add negative feedback to an ideal op amp (from this point on we assume ideal op amps) as shown in Fig. 6.2. It is easy to analyze how such a circuit will behave if you follow the following *Golden Rules of Op Amps*:

1. The output attempts to do whatever is necessary to make the voltage difference between the inputs zero.

2. The inputs draw no current.

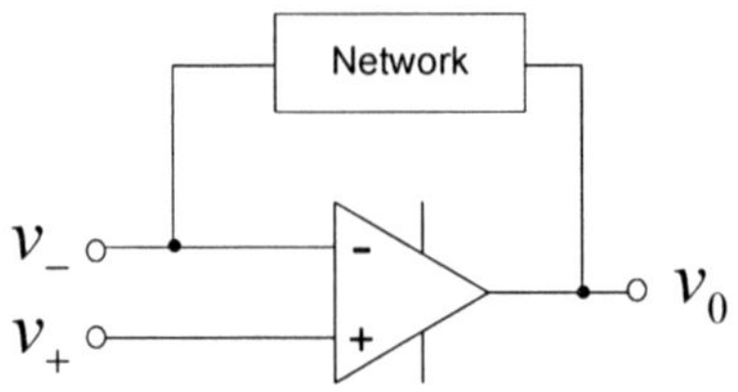

Figure 6.2 An ideal op amp with negative feedback

Note that the feedback network is always connected to the –, or inverting input to the op amp. We will give several example circuits of using such negative feedback in the remaining sections of this chapter.

6.2 Non-inverting Amplifier

An op amp circuit for a non-inverting amplifier is shown in Fig. 6.3. Note that we have swapped the + and – inputs from that shown in Fig. 6.2. The closed-loop voltage gain is, by definition,

$$A_F = \frac{v_o}{v_i} \tag{6.1}$$

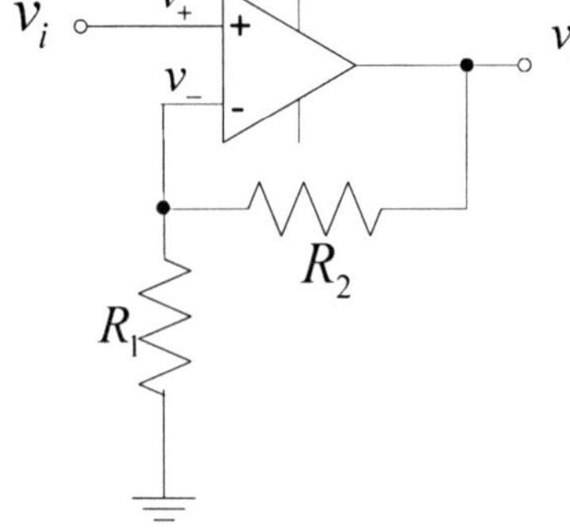

Note that resistors R_1 and R_2 form a voltage divider. Using this fact together with golden rule 1, we can write

$$v_i = v_+ = v_- = \frac{R_1}{R_1 + R_2} v_o \tag{6.2}$$

Figure 6.3 Non-inverting amplifier

Combining Eqs. (6.1) and (6.2), we see that the closed-loop gain of this non-inverting amplifier is given by

$$A_F = \frac{v_o}{v_i} = \frac{R_1 + R_2}{R_1} = 1 + \frac{R_2}{R_1} \tag{6.3}$$

Note that, if $R_2 = 0$, the closed-loop gain is 1 regardless of the value of R_1. Therefore, we don't need R_1 and, in this case, the circuit reduces to that shown in Fig. 6.4 and is called a unity-gain buffer. While this circuit provides no voltage gain, it is used as a "line driver" that transforms a high input impedance (resistance) to a low output impedance. This can provide substantial current gain.

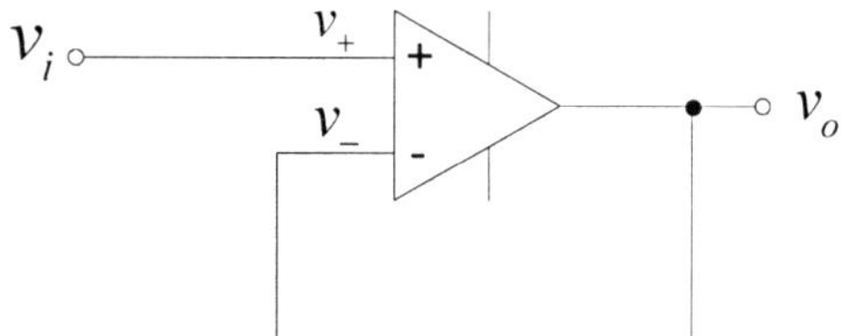

Figure 6.4 Unity-gain buffer

Example 29 – Non-inverting Amplifier

In the circuit shown in Fig. 6.5, let $R_1 = 2$ kΩ. What value of resistance R_2 will produce a voltage gain of 13?

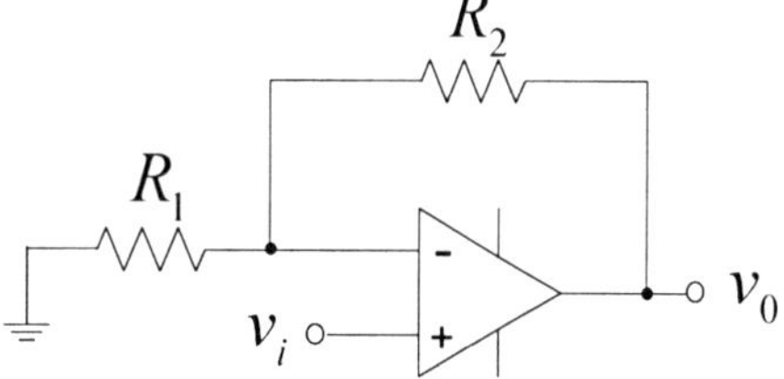

Figure 6.5 Example 29

This is really the same circuit as in Fig. 6.3 and R_1 and R_2 produce a voltage divider. Therefore, Eqs. (6.1) – (6.3) apply and we can solve Eq. (6.3) for R_2 to obtain

$$R_2 = R_1 \left(A_f - 1 \right) \tag{6.4}$$

Therefore, $R_2 = 24$ kΩ as shown in Matlab Example 29.

Matlab Example 29

```
>> r1 = 2;
>> Af = 13;
>> r2 = r1*(Af - 1)
r2 =
    24
```

6.3 Inverting Amplifier

An op amp circuit for an inverting amplifier is shown in Fig. 6.6. Applying golden rule 1, we see that

$$v_- = v_+ = 0 \qquad (6.5)$$

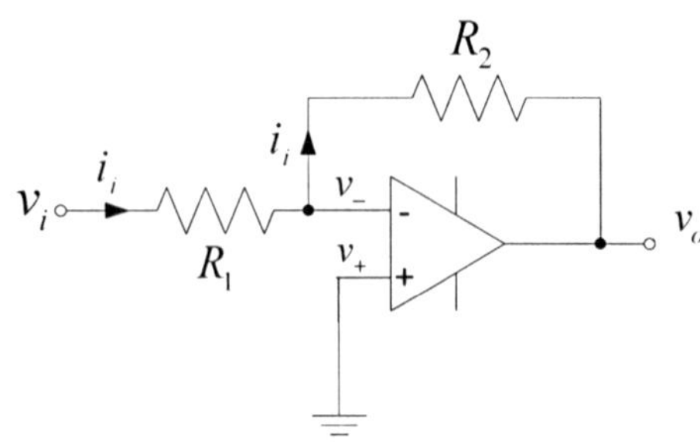

By golden rule 2, no current enters the op-amp inputs so that the current through the R_1 resistor will be equal to the current through the R_2 resistor as shown in Fig. 6.6. The current through the R_1 resistor is

$$i_i = \frac{v_i - 0}{R_1} = \frac{v_i}{R_1} \qquad (6.6)$$

Figure 6.6 Inverting amplifier

The current through the R_2 resistor is

$$i_i = \frac{0 - v_0}{R_2} = \frac{-v_0}{R_2} \qquad (6.7)$$

Equating Eqs. (6.6) and (6.7), we obtain

$$\frac{v_i}{R_1} = \frac{-v_0}{R_2}$$

from which the closed-loop gain is

$$A_F = \frac{v_o}{v_i} = -\frac{R_2}{R_1} \qquad (6.8)$$

Example 30 – Inverting Amplifier

In the circuit shown in Fig. 6.7, find the output voltage v_0.

From golden rule 1, the voltages at a and b are equal and from golden rule 2 no current enters the op amp. Therefore, if we apply KCL to node a, we can write

$$\frac{6 \text{ V} - 4 \text{ V}}{10 \text{ k}\Omega} = \frac{4 \text{ V} - v_0}{40 \text{ k}\Omega}$$

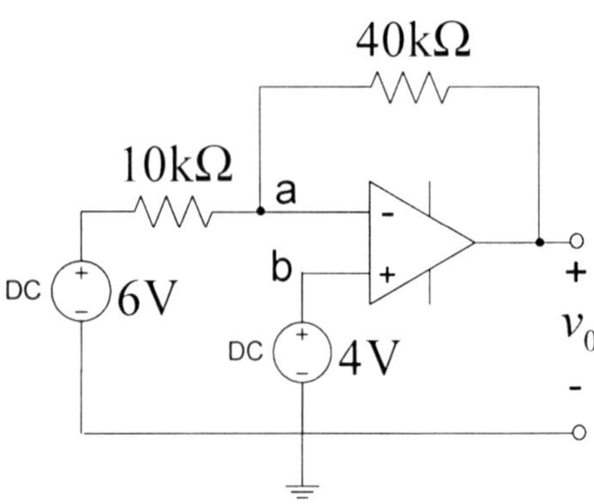

Figure 6.7 Example 23

or

$$4(6\text{ V} - 4\text{ V}) = 4\text{ V} - v_0$$

from which

$$v_0 = 4\text{ V} - 4(6\text{ V} - 4\text{ V}) = -4\text{ V}$$

Matlab Example 30

```
>> v0 = 4 - 4*(6 - 4)
V0 =
      -4
```

6.4 Differential Amplifier

The circuit shown in Fig. 6.8 is called a differential amplifier. There are two input voltages v_1 and v_2 and the output voltage v_0 is proportional to the difference of the two input voltages as we shall see.

From our two golden rules, we know that

$$v_- = v_+ \qquad (6.9)$$

and the current i_1 from v_1 to v_- is equal to the current i_1 from v_- to v_0. The current i_1 from v_1 to v_- is given by

$$i_1 = \frac{v_1 - v_-}{R_1} \qquad (6.10)$$

and the current i_1 from v_- to v_0 is given by

$$i_1 = \frac{v_- - v_0}{R_2} \qquad (6.11)$$

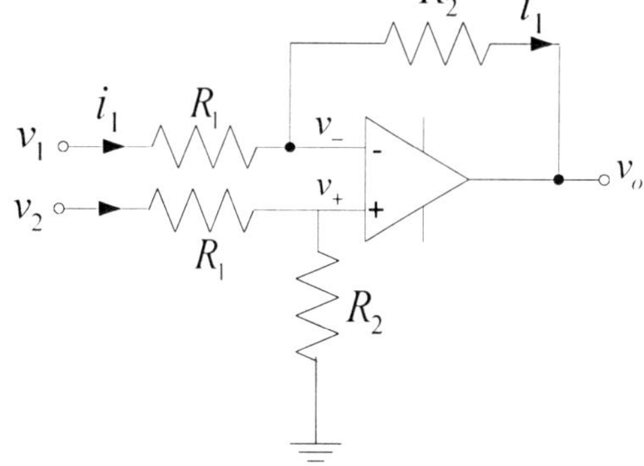

Figure 6.8 Differential amplifier

Equating Eqs. (6.10) and (6.11) and using Eq. (6.9), we obtain

$$\frac{v_1 - v_+}{R_1} = \frac{v_+ - v_0}{R_2} \qquad (6.12)$$

The two resistors connected to v_+ form a voltage divider so we can write

$$v_+ = \frac{R_2}{R_1 + R_2} v_2 \qquad (6.13)$$

Substituting Eq. (6.13) into Eq. (6.12) we obtain

$$\frac{v_1 - \dfrac{R_2}{R_1 + R_2}v_2}{R_1} = \frac{\dfrac{R_2}{R_1 + R_2}v_2 - v_0}{R_2} \tag{6.14}$$

from which

$$v_0 = -\frac{R_2}{R_1}v_1 + \frac{R_2}{R_1 + R_2}v_2 + \frac{R_2^2}{R_1(R_1 + R_2)}v_2$$

$$v_0 = -\frac{R_2}{R_1}v_1 + \frac{R_2}{R_1 + R_2}\left(1 + \frac{R_2}{R_1}\right)v_2$$

or

$$v_0 = \frac{R_2}{R_1}(v_2 - v_1) \tag{6.15}$$

Thus, the output of this differential amplifier is proportional to $(v_2 - v_1)$ and the voltage gain is R_2/R_1. For this circuit to work properly, it is necessary for both R_1 resistors and both R_2 resistors to have exactly the same values. It is actually the ratio of the two upper resistors and two lower resistors in Fig. 6.8 that is important (see Problem 6.1). Another problem with the circuit in Fig. 6.8 is that the input resistances seen by the two input voltages are no longer very large. This could cause the differential amplifier to load down the sensor network providing the two input voltages. A possible solution to this problem is to add a unity gain buffer of the kind shown in Fig. 6.4 to each of the input voltage lines. A better solution is to use an instrumentation amplifier such as the one described next in Example 32.

Example 31 – Differential Amplifier

Find the output voltage $v0$ in the differential amplifier circuit shown in Fig. 6.9.

From our two golden rules, we know that

$$v_- = v_+ \tag{6.16}$$

and the current i_1 through the top 2 kΩ resistor is equal to the current i_1 through the 30 kΩ resistor. Equating these two currents and using Eq. (6.16), we obtain

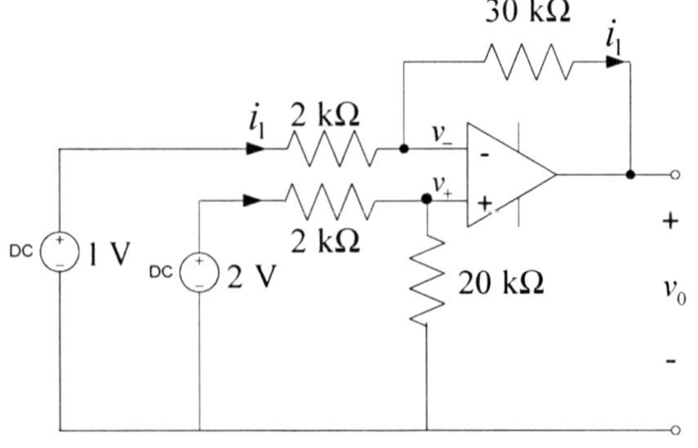

Figure 6.9 Circuit for Example 31

$$\frac{1\text{ V}-v_+}{2\text{ k}\Omega}=\frac{v_+-v_0}{30\text{ k}\Omega} \tag{6.17}$$

from which

$$v_0 = 16v_+ - 15\text{ V} \tag{6.18}$$

The two resistors connected to v_+ form a voltage divider so we can write

$$v_+ = \frac{20\text{ k}\Omega}{2\text{ k}\Omega+20\text{ k}\Omega}\,2\text{ V}=1.818\text{ V} \tag{6.19}$$

Substituting Eq. (6.19) into Eq. (6.18), we obtain

$$v_0 = 16(1.818)-15=14.09\text{ V} \tag{6.20}$$

Example 32 – Instrumentation Amplifier

Find the gain of the instrumentation amplifier shown in Fig. 6.10.

Op amp 3 is just a differential amplifier of the kind shown in Fig. 6.8 with a gain of 1. Thus, from Eq. (6.15),

$$v_0 = (v_4 - v_3) \tag{6.21}$$

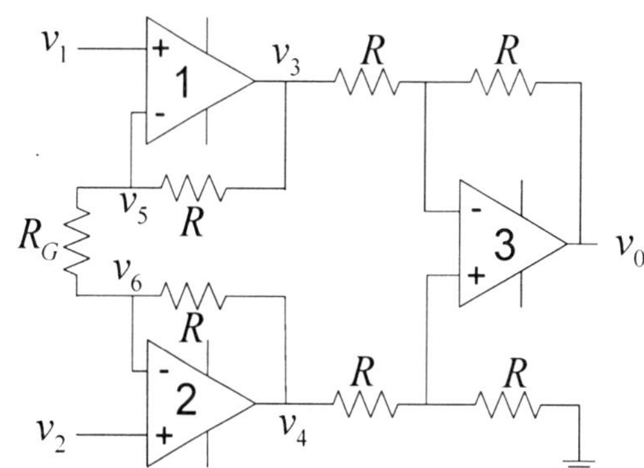

Figure 6.10 Instrumentation amplifier

From our golden rule 2, no current enters the $-$ inputs of op amps 1 and 2 in Fig. 6.10. Therefore, the current flowing through the three resistors along the path $v_3 - v_5 - v_6 - v_4$ is

$$i = \frac{(v_3 - v_4)}{2R + R_G} \tag{6.22}$$

This current can also be written as

$$i = \frac{(v_5 - v_6)}{R_G}=\frac{(v_1 - v_2)}{R_G} \tag{6.23}$$

where golden rule 1 was used in the last step. Equating Eqs. (6.22) and (6.23) after substituting Eq. (6.21) into Eq. (6.22), we obtain

$$\frac{-v_0}{2R+R_G} = \frac{\left(v_1-v_2\right)}{R_G}$$

from which

$$v_0 = \frac{2R+R_G}{R_G}\left(v_2-v_1\right) = \left(1+\frac{2R}{R_G}\right)\left(v_2-v_1\right) \qquad (6.24)$$

Therefore, the output voltage v_0 is proportional to the difference of the two input voltages with a gain given by

$$A_F = \left(1+\frac{2R}{R_G}\right) \qquad (6.25)$$

Instrumentation amplifiers similar to Fig. 6.10 are available in a single package where the only external resistor is R_G which controls the gain according to Eq. (6.25).

Matlab Example 32 calculates Eq. (6.25) and plots the result in Fig. 6.11 for a value of $R = 10$ kΩ. Note that the plot in Fig. 6.11 is a log-log plot. You can change the axes on your Matlab plots to log scale by selecting

Edit -> Axes Properties…

and then selecting the log scale as shown in Fig. 6.12. Use the *Insert* tab to add X and Y labels and a title to the plot.

Matlab Example 32

```
>> r = 10000;
>> rg = 100:100:10000;
>> Af = 1 + 2.*r./rg;
>> plot(rg,Af)
```

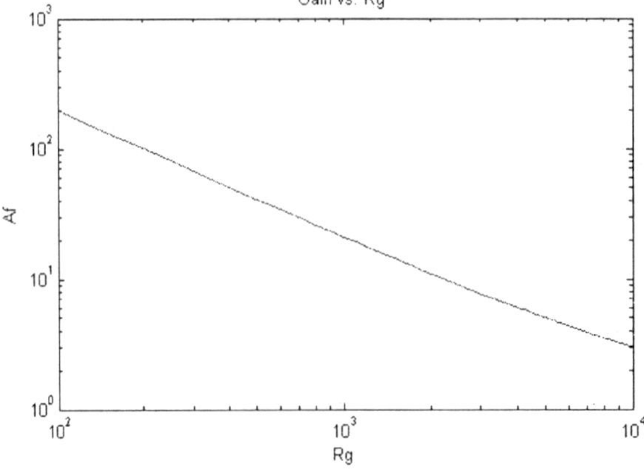

Figure 6.11 Plot of instrumentation amplifier gain

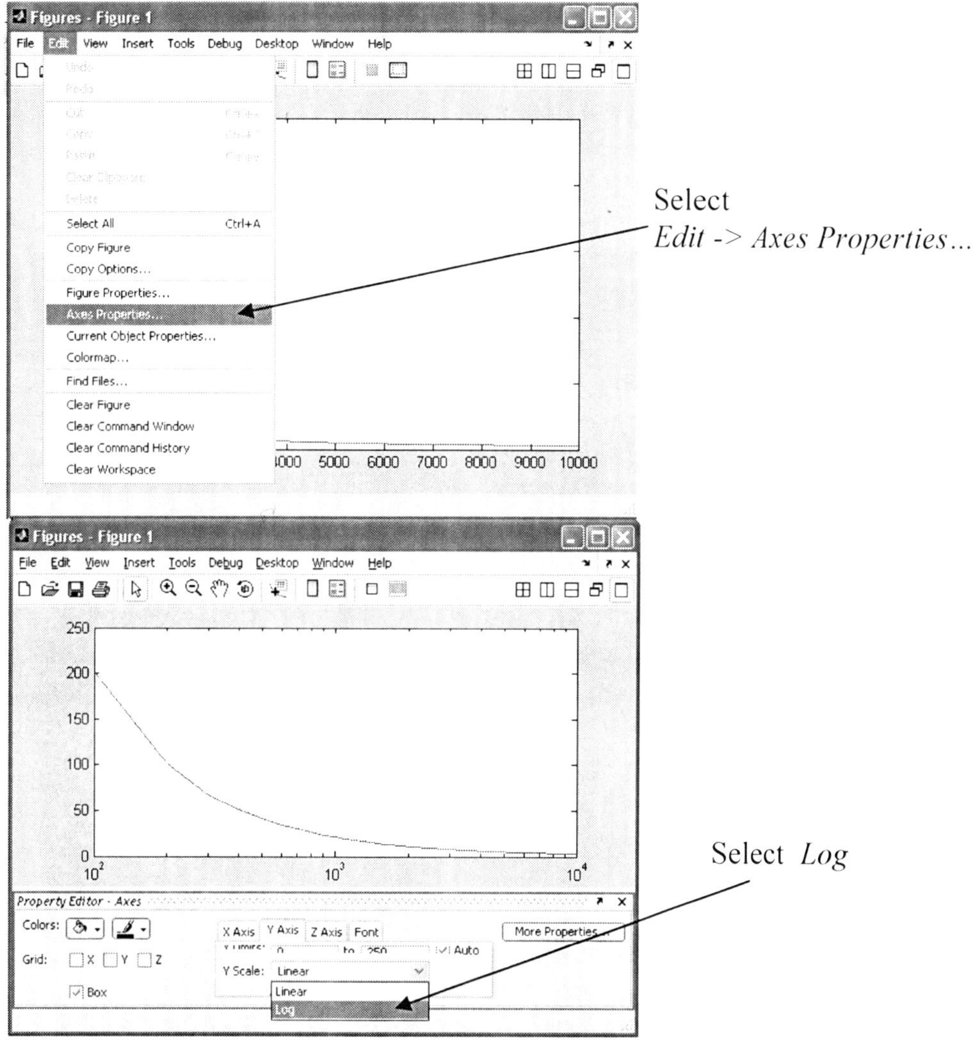

Figure 6.12 Changing the scale to a log plot

6.5 Active Filters

Suppose we replace the resistors in the circuit for an inverting amplifier in Fig. 6.6 with impedances that could contain a capacitor. The resulting circuit is shown in Fig. 6.13. The voltages and currents are now phasors representing sinusoidal AC signals. Applying golden rule 1, we see that

$$\mathbf{V}_- = \mathbf{V}_+ = 0 \qquad (6.26)$$

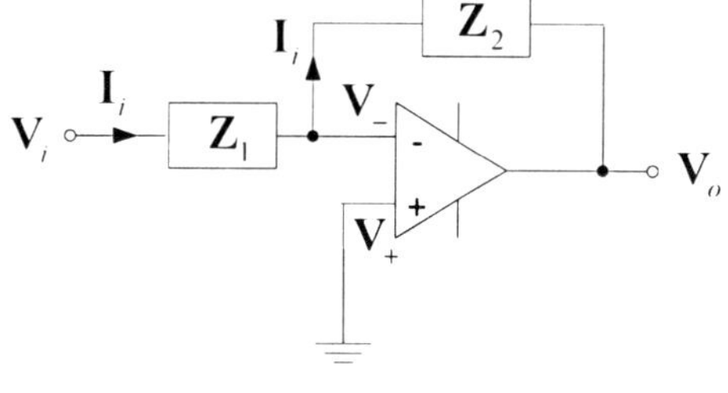

Figure 6.13
Inverting amplifier with impedances

By golden rule 2, no current enters the op-amp inputs so that the current through the $\mathbf{Z}_1$ impedance will be equal to the current through the $\mathbf{Z}_2$ impedance as shown in Fig. 6.13. The current through the $\mathbf{Z}_1$ impedance is

$$\mathbf{I}_i = \frac{\mathbf{V}_i - 0}{\mathbf{Z}_1} = \frac{\mathbf{V}_i}{\mathbf{Z}_1} \tag{6.27}$$

The current through the $\mathbf{Z}_2$ impedance is

$$\mathbf{I}_i = \frac{0 - \mathbf{V}_0}{\mathbf{Z}_2} = \frac{-\mathbf{V}_0}{\mathbf{Z}_2} \tag{6.28}$$

Equating Eqs. (6.27) and (6.28), we obtain

$$\frac{\mathbf{V}_i}{\mathbf{Z}_1} = \frac{-\mathbf{V}_0}{\mathbf{Z}_2} \tag{6.29}$$

We define the *transfer function* $\mathbf{H}(\omega)$ as the ratio of the output phasor voltage to the input phasor voltage. Note that it will be a complex number and from Eq. (6.29) is given by

$$\mathbf{H}(\omega) = \frac{\mathbf{V}_0}{\mathbf{V}_i} = \frac{-\mathbf{Z}_2}{\mathbf{Z}_1} \tag{6.30}$$

Example 33 – Low-Pass Filter

In the circuit shown in Fig. 6.14, if $R_1 = 200\ \Omega$, $R_2 = 2\ \text{k}\Omega$, and $C = 10\ \text{nF}$, plot the magnitude of the transfer function $\mathbf{H}(\omega)$ for frequencies between 10 Hz and 100 kHz.

Comparing this circuit to that in Fig. 6.13, we can write

$$\mathbf{Z}_1 = R_1 \tag{6.31}$$

and

$$\mathbf{Z}_2 = \frac{R_2 / j\omega C}{R_2 + \dfrac{1}{j\omega C}} = \frac{R_2}{1 + j\omega R_2 C} \tag{6.32}$$

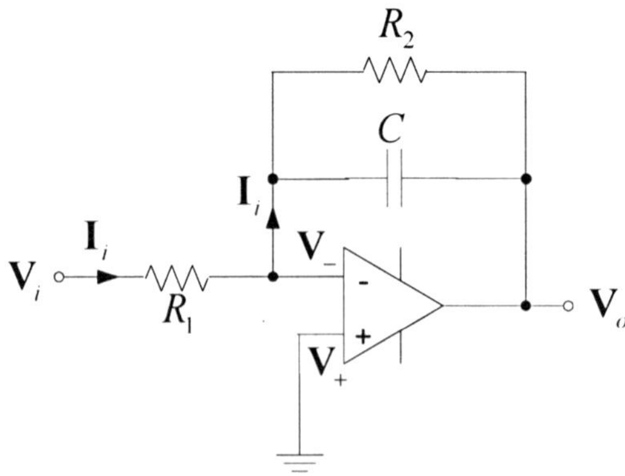

Figure 6.14 Active low-pass filter

Substituting Eqs. (6.31) and (6.32) into Eq. (6.30), we obtain

$$\mathbf{H}(\omega) = \frac{-\mathbf{Z}_2}{\mathbf{Z}_1} = -\frac{R_2}{R_1}\frac{1}{1+j\omega R_2 C} \tag{6.33}$$

Matlab Example 33 will produce the plot shown in Fig. 6.15. Note that the gain is about $R_2/R_1 = 10$ for frequencies less than 1 kHz and then drops off to about half by 10 kHz.

Matlab Example 33

```
>> R1 = 200;
>> R2 = 2000;
>> C = 10*10^-9;
>> f = linspace(10,100000,100);
>> w = 2*pi*f;
>> H = -(R2/R1).*(1./(1+j.*w*R2*C));
>> absH = abs(H);
>> plot(f,absH)
```

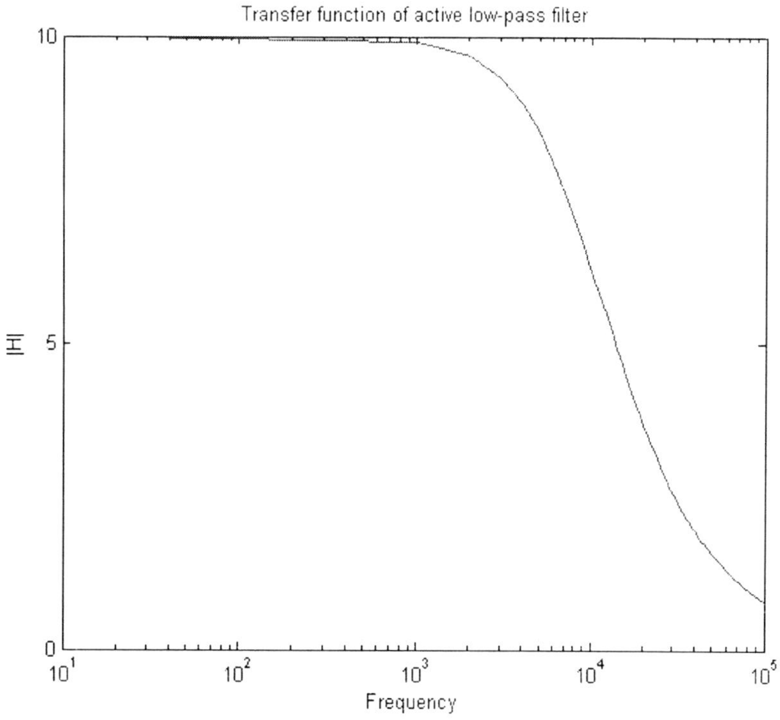

Figure 6.15 Plot resulting from Matlab Example 33

Example 34 – High-Pass Filter

In the circuit shown in Fig. 6.16, if $R_1 = 5\text{ k}\Omega$, $R_2 = 20\text{ k}\Omega$, and $C = 20\text{ nF}$, plot the magnitude of the transfer function $\mathbf{H}(\omega)$ for frequencies between 10 Hz and 1 MHz.

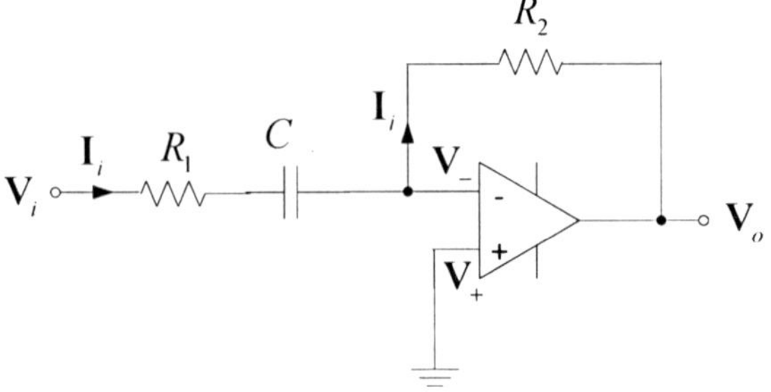

Figure 6.16 Active high-pass filter

Comparing this circuit to that is Fig. 6.13, we can write

$$\mathbf{Z}_1 = R_1 + 1/j\omega C \qquad (6.34)$$

and

$$\mathbf{Z}_2 = R_2 \qquad (6.35)$$

Substituting Eqs. (6.34) and (6.35) into Eq. (6.30), we obtain

$$\mathbf{H}(\omega) = \frac{-\mathbf{Z}_2}{\mathbf{Z}_1} = -\frac{R_2}{R_1 + 1/j\omega C} = -\frac{j\omega R_2 C}{1 + j\omega R_1 C} \qquad (6.36)$$

Matlab Example 34 will produce the plot shown in Fig. 6.17. Note that the gain increases linearly at low frequencies and approaches a maximum gain of $R_2/R_1 = 4$ for frequencies greater than 10 kHz.

Matlab Example 34

```
>> R1 = 5000;
>> R2 = 20000;
>> C = 20*10^-9;
>> f = linspace(10,1000000,100);
>> w = 2*pi.*f;
>> H = (j.*w*R2*C)./(1+j.*w*R1*C);
>> absH = abs(H);
>> plot(f,absH)
```

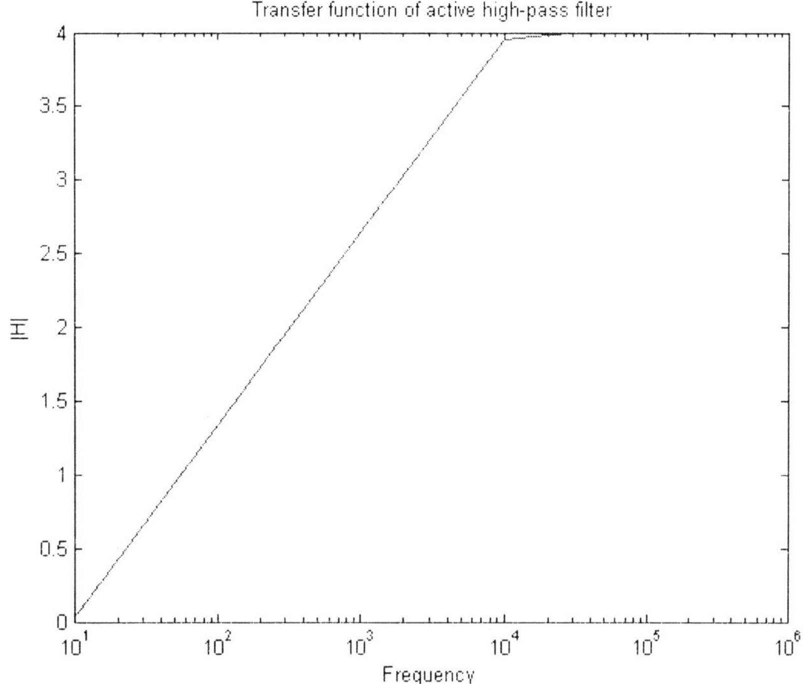

Figure 6.17 Plot resulting from Matlab Example 34

6.6 Current-to-Voltage Converter

The circuit shown in Fig. 6.18 will convert an input current i_i to an output voltage v_0. From golden rule 2,

$$i_i = i_f \qquad (6.37)$$

and from golden rule 1,

$$v_- = v_+ = 0 \qquad (6.38)$$

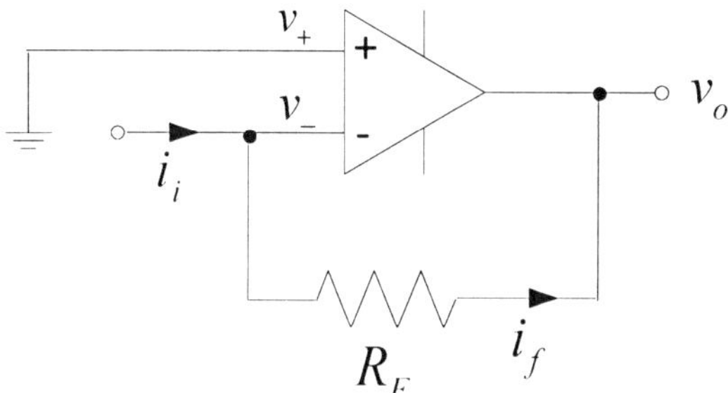

Figure 6.18 Current-to-voltage converter

We can therefore write the voltage across resistor R_F as

$$0 - v_0 = i_f R_F \qquad (6.39)$$

from which, using Eq. (6.37),

$$v_0 = -i_i R_F \qquad (6.40)$$

The *transresistance* is defined as the change in output voltage per unit change in input current. From Eq. (6.40), we see that the transresistance for the circuit in Fig. 6.18 is

$$\text{Transresistance} = \Delta v_0 / \Delta i_i = -R_F \qquad (6.41)$$

Example 35 – Photodiode Circuit

The circuit shown in Fig. 6.13 can be used to measure the amount of light incident on a photodiode. The reverse light current i_i is proportional to the incident light radiation. Assume that this light current is 25 μA per mW/cm^2 of incident radiation. If the incident radiation is 50 mW/cm^2, find the resistance R_F that will produce an output voltage of –4V.

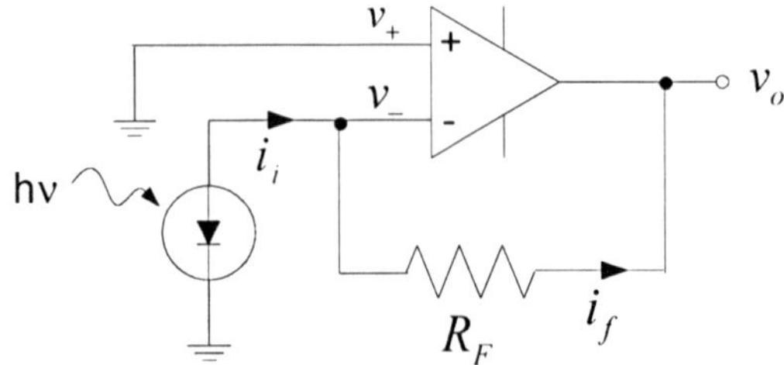

Figure 6.13 Current-to-voltage converter

From Matlab Example 35, we see that the current from the photodiode will be

$$i_i = 50 \frac{\text{mW}}{\text{cm}^2} \times 25 \frac{\mu A}{\text{mW}/\text{cm}^2} = 1.25 \text{ mA} \qquad (6.42)$$

From Eqs. (6.40) and (6.42), we see that

$$R_F = -\frac{v_0}{i_i} = -\frac{-4 \text{ V}}{1.25 \text{ mA}} = 3.2 \text{ k}\Omega \qquad (6.43)$$

Matlab Example 35

```
>> v0 = -4;
>> il = 25*10^-6;
>> rad = 50;
>> ii = il * rad
ii =
   1.2500e-003

>> rf = -v0/ii
rf =
   3.2000e+003
```

Problems

6.1   In the circuit to the right, show that if

$$\frac{R_1}{R_2} = \frac{R_3}{R_4}$$

then

$$v_0 = \frac{R_2}{R_1}\left(v_2 - v_1\right)$$

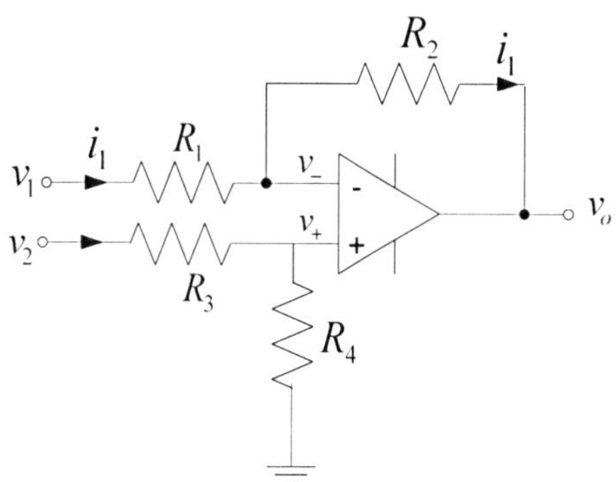

6.2   In the circuit shown below, find the value of the output voltage v_0.

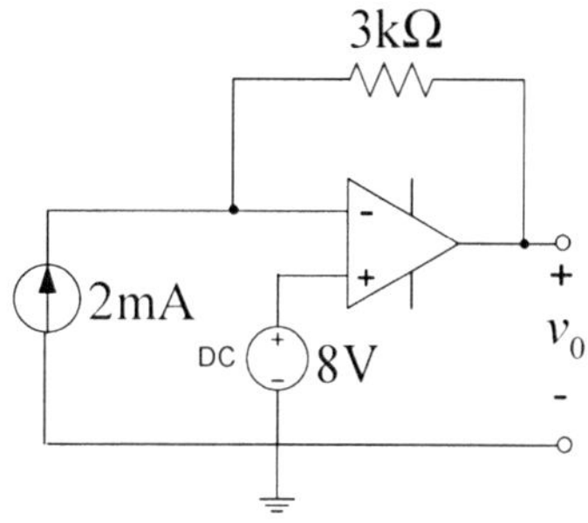

6.3   In the circuit shown below, find the value of the output voltage v_0.

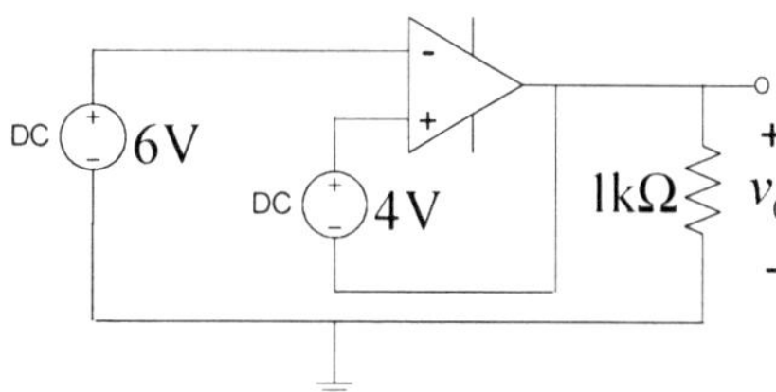

6.4.   Find the output voltage v_0 in the following ideal op amp circuit.

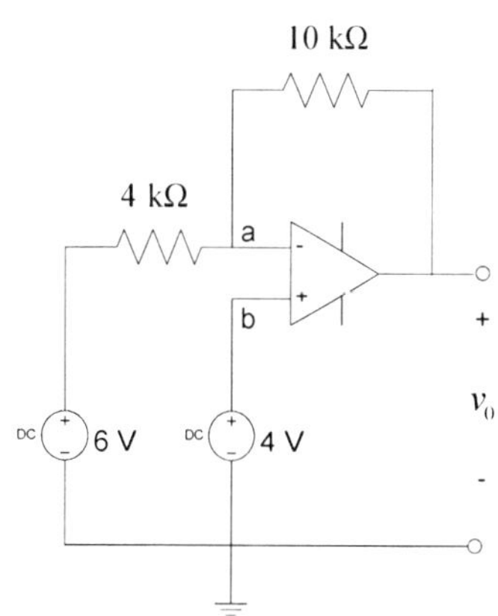

6.5. Find the current i_0 and the voltage v_0 in the following ideal op amp circuit.

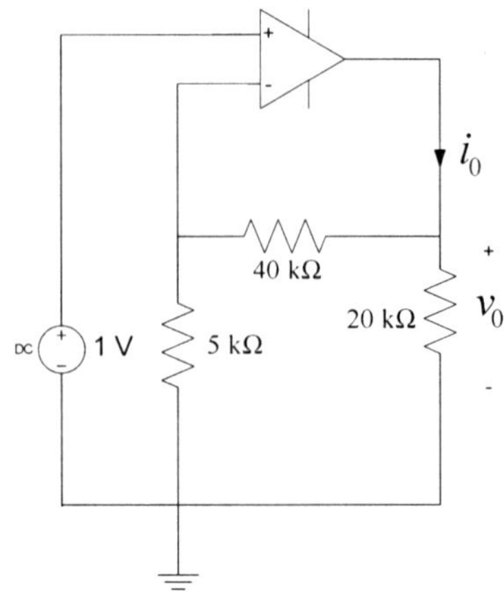

6.6. In the following circuit, if $\mathbf{V}_i = 10\angle 0^\circ$ V, $R_1 = 200\ \Omega$, $R_2 = 2$ kΩ, and $C = 1$ nF, find the output phasor voltage $\mathbf{V}_o$ at a frequency of 10 kHz.

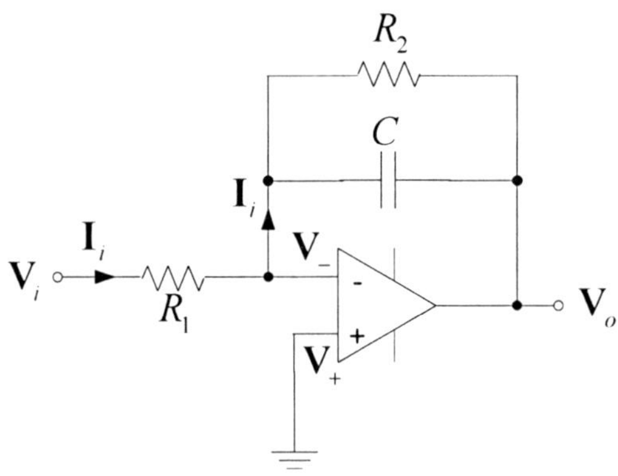

6.7. In the op amp circuit shown below, find the closed loop gain A_F if $R_1 = 3$ kΩ and $R_2 = 21$ kΩ.

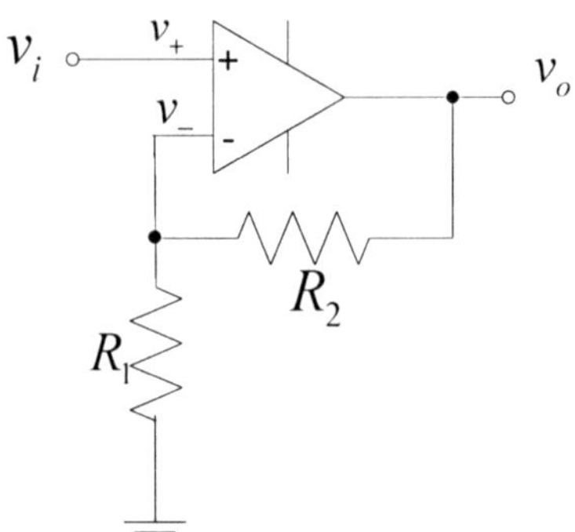

Chapter 7

Mesh and Nodal Analysis

The goal of circuit analysis is to be able to calculate all the currents and/or all the voltages in a given circuit. In this chapter, we will present two general methods, *mesh analysis* and *nodal analysis*, which can be used to solve any planar circuit problem involving resistors, capacitors, or inductors, and independent voltage and current sources. Both methods represent systematic ways of using Kirchoff's laws and Ohm's Law to write equations that can be solved easily using Matlab or a calculator to yield a set of variables that make calculating all the currents and voltages in the circuit very easy. The choice of which method to use for a given circuit depends on the number of equations required by each method, since the fewer the equations, the less time it takes to write them. Thus, unless the problem specifies a method, we will always choose the method that requires the fewest equations. We will show how you can write mesh and nodal equations by inspection and how you can solve circuits that contain both voltage and current sources.

7.1 Mesh Analysis

In *mesh analysis*, we apply Kirchhoff's voltage law (KVL) to find unknown currents. It is applicable only to *planar* circuits (a circuit that can be drawn on a plane with no branches crossing each other). Recall from Chapter 3 that a *mesh* is a loop that does not contain any other loops, that there are l meshes in a circuit, and that using the Fundamental Theorem of Circuit Analysis l can be found by counting the number of nodes and the number of elements in the circuit. Since a mesh is a closed path in the circuit, a current can flow in the mesh. The current through a mesh is known as the *mesh current*. In this section, we will assume that the circuit contains only voltage sources. Thus, we are assuming that any current sources in the circuit can be exchanged for voltage sources (see Section 5.4). We will treat the case of circuits containing both current and voltage sources in Section 7.5.

The following steps are used for mesh analysis.

1. Assign mesh currents i_1, i_2, i_3, ... i_l, to the l meshes, and make sure all these currents flow in the same direction,
2. Apply KVL to each of the l meshes and use Ohm's law to express the voltages across each resistor in terms of the mesh currents,
3. Solve the l resulting simultaneous equations to find the mesh currents.

Example 36 – Mesh Analysis

We will illustrate the use of mesh analysis using the circuit shown in Fig. 7.1. The first step is to assign the three mesh currents i_1, i_2, and i_3. The direction of these mesh currents is arbitrary but they must all have the same direction. We have drawn them clockwise in Fig. 7.1. Note that the mesh currents are not necessarily real currents in the circuit. The mesh current i_1 is the actual current flowing through resistor R_1. However, the real current flowing through resistor R_3 (from the + side of v_3) is $i_3 - i_1$. In mesh analysis we solve for the mesh currents and then use these mesh currents to calculate the actual branch currents and the voltages across each resistor.

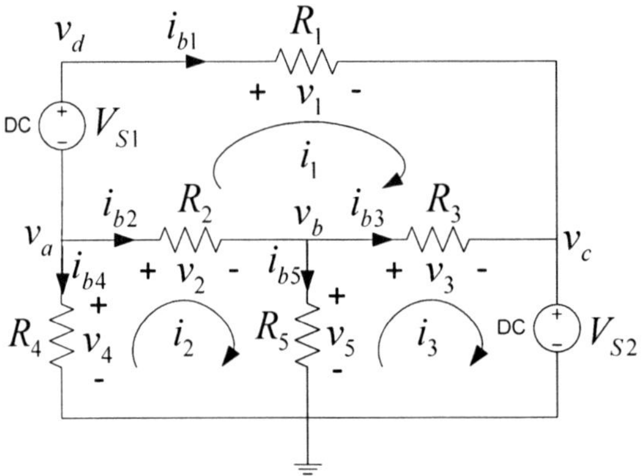

Figure 7.1 Illustrating mesh analysis

The next step is to apply KVL to each of the three meshes. We will sum the voltages around each loop taking the sign of the voltage to be the first sign encountered in going around the loop.

$$\text{Mesh 1:} \quad -V_{S1} + v_1 - v_3 - v_2 = 0 \tag{7.1}$$

$$\text{Mesh 2:} \quad -v_4 + v_2 + v_5 = 0 \tag{7.2}$$

$$\text{Mesh 3:} \quad -v_5 + v_3 + V_{S2} = 0 \tag{7.3}$$

We now use Ohm's law to express the voltages in Eqs. (7.1) – (7.3) in terms of the mesh currents.

$$\text{Mesh 1:} \quad -V_{S1} + i_1 R_1 - \left(i_3 - i_1\right) R_3 - \left(i_2 - i_1\right) R_2 = 0 \tag{7.4}$$

$$\text{Mesh 2:} \quad -\left(-i_2 R_4\right) + \left(i_2 - i_1\right) R_2 + \left(i_2 - i_3\right) R_5 = 0 \tag{7.5}$$

$$\text{Mesh 3:} \quad -\left(i_2 - i_3\right) R_5 + \left(i_3 - i_1\right) R_3 + V_{S2} = 0 \tag{7.6}$$

The next step is to rearrange the terms in Eqs. (7.4) – (7.6) by collecting all of the coefficients of the mesh currents.

$$\text{Mesh 1: } \left(R_1 + R_3 + R_2\right)i_1 - R_2 i_2 - R_3 i_3 = V_{S1} \qquad (7.7)$$

$$\text{Mesh 2: } -R_2 i_1 + \left(R_4 + R_2 + R_5\right)i_2 - R_5 i_3 = 0 \qquad (7.8)$$

$$\text{Mesh 3: } -R_3 i_1 - R_5 i_2 + \left(R_5 + R_3\right)i_3 = -V_{S2} \qquad (7.9)$$

The set of three linear equations given by Eqs. (7.7) – (7.9) can be written in matrix form as

$$\begin{bmatrix} R_1 + R_3 + R_2 & -R_2 & -R_3 \\ -R_2 & R_4 + R_2 + R_5 & -R_5 \\ -R_3 & -R_5 & R_5 + R_3 \end{bmatrix} \begin{bmatrix} i_1 \\ i_2 \\ i_3 \end{bmatrix} = \begin{bmatrix} V_{S1} \\ 0 \\ -V_{S2} \end{bmatrix} \qquad (7.10)$$

As a specific example, let $V_{S1} = 4\text{V}$, $V_{S2} = 2\text{V}$, $R_1 = 2\Omega$, $R_2 = 3\Omega$, $R_3 = 4\Omega$, $R_4 = 1\Omega$, and $R_5 = 2\Omega$. Then Matlab Example 36 shows that the three mesh currents are $i_1 = 0.7556\text{A}$, $i_2 = 0.4889\text{A}$, and $i_3 = 0.3333\text{A}$. The branch currents can then be found in terms of the mesh currents as follows.

$$i_{b1} = i_1 = 0.7556\text{A}$$

$$i_{b2} = i_2 - i_1 = -0.2667\text{A}$$

$$i_{b3} = i_3 - i_1 = -0.4222\text{A}$$

$$i_{b4} = -i_2 = -0.4889\text{A}$$

$$i_{b2} = i_2 - i_3 = 0.1556\text{A}$$

The three node voltages are then given by

$$v_a = i_{b4} R_4 = -0.4889\text{V}$$

$$v_b = i_{b5} R_5 = 0.3111\text{V}$$

$$v_c = V_{S2} = 2\text{V}$$

Matlab Example 36

```
>> Vs1 = 4;
>> Vs2 = 2;
>> r1 = 2;
>> r2 = 3;
>> r3 = 4;
>> r4 = 1;
>> r5 = 2;
>> R = [r1+r3+r2 -r2 -r3; -r2 r4+r2+r5 -r5; -r3 -r5 r5+r3]
R =
     9    -3    -4
    -3     6    -2
    -4    -2     6
```

Matlab Example 36 (cont.)

```
>> v = [Vs1 0 -Vs2]'
v =
     4
     0
    -2

>> i = R\v
i =
    0.7556
    0.4889
    0.3333

>> ib1 = i(1)
ib1 =
    0.7556

>> ib2 = i(2)-i(1)
ib2 =
   -0.2667

>> ib3 = i(3)-i(1)
ib3 =
   -0.4222

>> ib4 = -i(2)
ib4 =
   -0.4889

>> ib5 = i(2)-i(3)
ib5 =
    0.1556

>> va = ib4*r4
va =
   -0.4889

>> vb = ib5*r5
vb =
    0.3111

>> vc = Vs2
vc =
     2
```

7.2 Mesh Analysis by Inspection

If you study Example 36 carefully, you will see that the matrix equation given by Eq. (7.10) can be written by inspection without having to go through all of the steps in Eqs. (7.1) – (7.9)!

Eq. (7.10) is of the form

$$\mathbf{Ri} = \mathbf{v} \qquad\qquad (7.11)$$

where $\mathbf{R}$ is an $l \times l$ symmetric resistance matrix, $\mathbf{i}$ is an $l \times 1$ vector of mesh currents, and $\mathbf{v}$ is a vector of voltages representing known voltages.

Here are the rules for writing Eq. (7.11) by inspection.

- The matrix $\mathbf{R}$ is symmetric, $r_{kj} = r_{jk}$ and all of the off-diagonal terms are negative or zero.
- The r_{kk} terms are the sum of all resistances in mesh k.
- The r_{kj} terms are the negative sum of the resistances common to BOTH mesh k and mesh j.
- The v_k (the kth component of the vector $\mathbf{v}$) are equal to the algebraic sum of the independent voltages in mesh k, with *voltage rises taken as positive and voltage drops taken as* negative when going around the mesh in the assumed direction of current flow.

We will illustrate these rules in Example 37.

Example 37 – Mesh Analysis by Inspection

Let's solve for the mesh currents in the circuit shown in Fig. 7.2 by inspection.

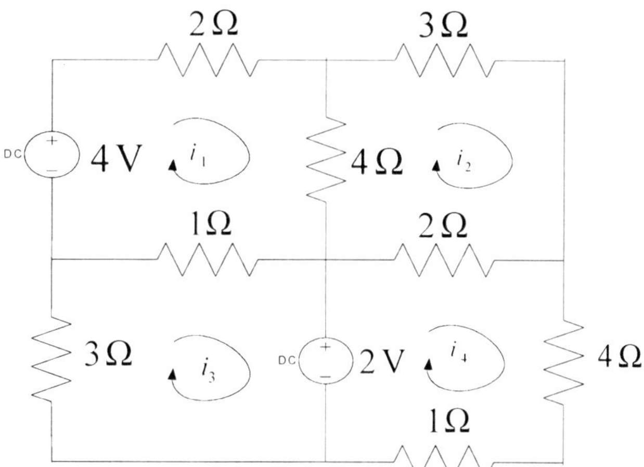

Figure 7.2 Illustrating mesh analysis by inspection

We begin by writing the matrix equation in the form

$$
\begin{bmatrix} \end{bmatrix}
\begin{bmatrix} i_1 \\ i_2 \\ i_3 \\ i_4 \end{bmatrix} =
\begin{bmatrix} \end{bmatrix}
\tag{7.12}
$$

Because we have four meshes, the resistance matrix will be 4 x 4 and the matrix of mesh currents will be 4 x 1. We'll begin by filling in the main diagonal of the resistance matrix by just adding the resistances around each mesh. This will give

$$\begin{pmatrix} 2+4+1 & & & \\ & 3+2+4 & & \\ & & 3+1 & \\ & & & 2+4+1 \end{pmatrix}\begin{pmatrix} i_1 \\ i_2 \\ i_3 \\ i_4 \end{pmatrix} = \begin{pmatrix} \\ \\ \\ \end{pmatrix} \tag{7.13}$$

Now we will fill in the off-diagonal terms. Remember, these will be the negative of the sum of the resistances that are common to two meshes. For example, a 4Ω resistor is common to meshes 1 and 2 so the R_{12} entry will be -4. The resistance matrix is symmetric so this will also be the entry for R_{21}. Filling in all the off-diagonal terms will give

$$\begin{pmatrix} 2+4+1 & -4 & -1 & 0 \\ -4 & 3+2+4 & 0 & -2 \\ -1 & 0 & 3+1 & 0 \\ 0 & -2 & 0 & 2+4+1 \end{pmatrix}\begin{pmatrix} i_1 \\ i_2 \\ i_3 \\ i_4 \end{pmatrix} = \begin{pmatrix} \\ \\ \\ \end{pmatrix} \tag{7.14}$$

We now only have to fill in the matrix of known voltages on the right-hand side of Eq. 7.14. These will be the sum of the independent voltage sources around each mesh, where a voltage *rise* is taken as positive. Remember, when we go around a mesh in the direction of mesh current flow and apply KVL, we take the first sign encountered as the sign of the voltage. In that case, a voltage rise will be negative when adding all the voltages around the mesh and setting the sum to zero. But now we have moved the independent voltage sources to the right-hand side of the equation so when using the method of inspection we must take voltage rises to be positive. The completed matrix equation will be

$$\begin{pmatrix} 2+4+1 & -4 & -1 & 0 \\ -4 & 3+2+4 & 0 & -2 \\ -1 & 0 & 3+1 & 0 \\ 0 & -2 & 0 & 2+4+1 \end{pmatrix}\begin{pmatrix} i_1 \\ i_2 \\ i_3 \\ i_4 \end{pmatrix} = \begin{pmatrix} 4 \\ 0 \\ -2 \\ 2 \end{pmatrix} \tag{7.15}$$

Matlab Example 37 shows the Matlab solution to Eq. (7.15). It doesn't get much easier than this! The alternative to using this method of inspection is to write equations similar to Eqs. (7.1) – (7.9) for all four meshes in Fig. 7.2 in order to derive Eq. (7.15). It is clearly easier to simply write Eq. (7.15) by inspection.

Matlab Example 37

```
>> R = [2+4+1 -4 -1 0; -4 3+2+4 0 -2; -1 0 3+1 0; 0 -2 0 2+4+1]
R =
      7     -4     -1      0
     -4      9      0     -2
     -1      0      4      0
      0     -2      0      7

>> v = [4 0 -2 2]'
v =
      4
      0
     -2
      2

>> i = R\v
i =
    0.7773
    0.4367
   -0.3057
    0.4105
```

Example 38 – AC Mesh Analysis by Inspection

In the circuit shown in Fig. 7.3, find the two phasor mesh currents and the phasor branch current $\mathbf{I}_0$.

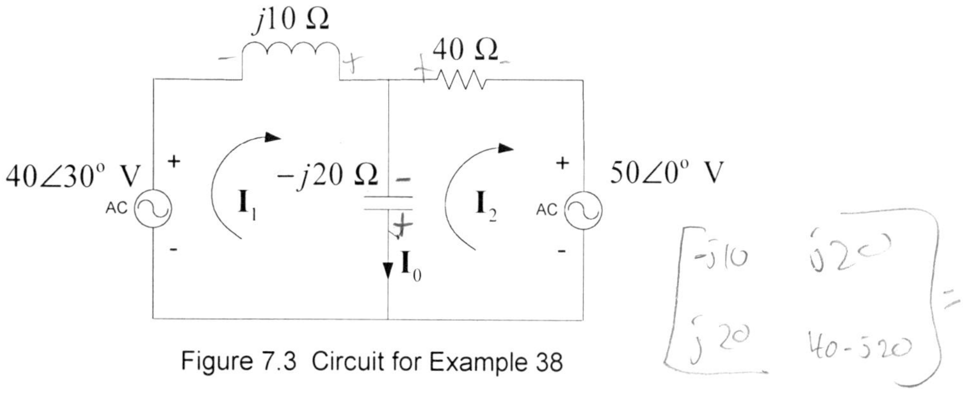

Figure 7.3 Circuit for Example 38

By inspection, we can write the matrix mesh equations as

$$\begin{bmatrix} -j10 & j20 \\ j20 & 40-j20 \end{bmatrix}\begin{bmatrix} \mathbf{I}_1 \\ \mathbf{I}_2 \end{bmatrix} = \begin{bmatrix} 40\angle30° \\ -50\angle0° \end{bmatrix} \tag{7.16}$$

From Matlab Example 38, we see that the solution is

$$\begin{bmatrix} \mathbf{I}_1 \\ \mathbf{I}_2 \end{bmatrix} = \begin{bmatrix} 4.698\angle 95.2^\circ \\ 0.993\angle 37.7^\circ \end{bmatrix} \qquad (7.17)$$

$$\mathbf{I}_0 = \mathbf{I}_1 - \mathbf{I}_2 = 4.249\angle 106.6^\circ \ \text{A} \qquad (7.18)$$

Matlab Example 38

```
>> Z = [-10j 20j; 20j 40-20j]
Z =
         0 -10.0000i          0 +20.0000i
         0 +20.0000i   40.0000 -20.0000i

>> V = [40*exp(j*30*pi/180); -50]
V =
  34.6410 +20.0000i
 -50.0000

>> I = Z\V
I =
  -0.4287 + 4.6785i
   0.7856 + 0.6072i

>> ampI = abs(I)
ampI =
    4.6981
    0.9929

>> angI = angle(I)*180/pi
angI =
   95.2358
   37.6985

>> I0 = I(1) - I(2)
I0 =
  -1.2144 + 4.0713i

>> magI0 = abs(I0)
magI0 =
    4.2485

>> angI0 = angle(I0)*180/pi
angI0 =
  106.6085
```

Note that we could *not* have written the column vector for **V** as

```
V = [40*exp(j*30*pi/180) -50]'
```

because taking the transpose of a complex row matrix will produce the complex conjugates of each element.

7.3 Nodal Analysis

In *nodal analysis*, we apply Kirchhoff's current law (KCL) to find unknown *node voltages with respect to a reference node*. In this section, we will assume that the circuit contains only current sources. Thus, we are assuming that any voltage sources in the circuit can be exchanged for current sources (see Section 5.4). We will treat the case of circuits containing both current and voltage sources in Section 7.6.

The following steps are used for nodal analysis.

1. Select one of the n nodes as a reference node (that we define to be zero voltage, or ground). Assign voltages v_1, v_2, ... v_{n-1} to the remaining n-1 nodes. These voltages are referenced with respect to the reference node.
2. Apply KCL to each of the n-1 non-reference nodes. Use Ohm's law to express the branch currents in terms of the node voltages.
3. Solve the resulting simultaneous equations to obtain the node voltages v_1, v_2, ... v_{n-1}

Example 39 – Nodal Analysis

We will illustrate the use of nodal analysis using the circuit shown in Fig. 7.4. The first step is to assign the three node voltages v_1, v_2, and v_3. In nodal analysis, we solve for the node voltages and then use these node voltages to calculate the voltages across each resistor and the branch currents.

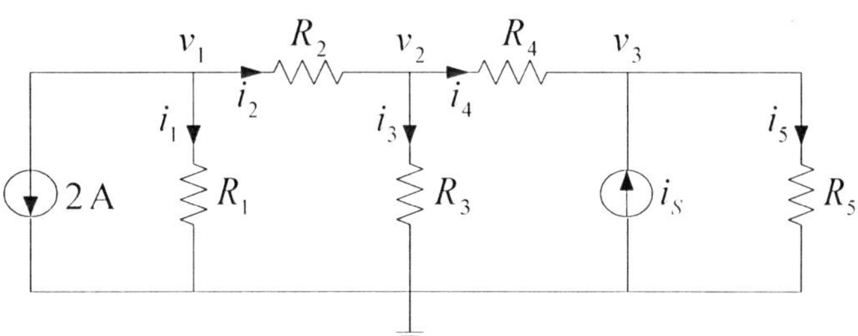

Figure 7.4 Illustrating nodal analysis

The next step is to apply KCL to each of the three nodes. We will sum the currents into each node taking currents leaving the node as positive.

$$\text{Node 1:} \quad 2 + i_1 + i_2 = 0 \tag{7.19}$$

$$\text{Node 2:} \quad -i_2 + i_3 + i_4 = 0 \tag{7.20}$$

$$\text{Node 3:} \quad -i_4 - i_s + i_5 = 0 \tag{7.21}$$

We now use Ohm's law to express the currents in Eqs. (7.19) – (7.21) in terms of the node voltages.

$$\text{Node 1: } 2+\frac{v_1}{R_1}+\frac{v_1-v_2}{R_2}=0 \tag{7.22}$$

$$\text{Node 2: } -\frac{v_1-v_2}{R_2}+\frac{v_2}{R_3}+\frac{v_2-v_3}{R_4}=0 \tag{7.23}$$

$$\text{Node 3: } -\frac{v_2-v_3}{R_4}-i_S+\frac{v_3}{R_5}=0 \tag{7.24}$$

The next step is to rearrange the terms in Eqs. (7.22) – (7.24) by collecting all of the coefficients of the node voltages. We need to make sure all node voltages are included in each equation (e.g. if one of the node voltages doesn't appear in the equation it enters with a zero coefficient).

$$\text{Node 1: } \left(\frac{1}{R_1}+\frac{1}{R_2}\right)v_1-\frac{1}{R_2}v_2+0v_3=-2 \tag{7.25}$$

$$\text{Node 2: } -\frac{1}{R_2}v_1+\left(\frac{1}{R_2}+\frac{1}{R_3}+\frac{1}{R_4}\right)v_2-\frac{1}{R_4}v_3=0 \tag{7.26}$$

$$\text{Node 3: } 0v_1-\frac{1}{R_4}v_2+\left(\frac{1}{R_4}+\frac{1}{R_5}\right)v_3=i_S \tag{7.27}$$

If we replace the resistances in Eqs. (7.25) – (7.27) with the corresponding conductances, we can write

$$\text{Node 1: } \left(G_1+G_2\right)v_1-G_2v_2+0v_3=-2 \tag{7.28}$$

$$\text{Node 2: } -G_2v_1+\left(G_2+G_3+G_4\right)v_2-G_4v_3=0 \tag{7.29}$$

$$\text{Node 3: } 0v_1-G_4v_2+\left(G_4+G_5\right)v_3=i_S \tag{7.30}$$

The set of three linear equations given by Eqs. (7.28) – (7.30) can be written in matrix form as

$$\begin{bmatrix} G_1+G_2 & -G_2 & 0 \\ -G_2 & G_2+G_3+G_4 & -G_4 \\ 0 & -G_4 & G_4+G_5 \end{bmatrix}\begin{bmatrix} v_1 \\ v_2 \\ v_3 \end{bmatrix}=\begin{bmatrix} -2 \\ 0 \\ i_S \end{bmatrix} \tag{7.31}$$

As a specific example, let $i_S=1\text{A}$, $R_1=2\Omega$, $R_2=1\Omega$, $R_3=4\Omega$, $R_4=3\Omega$, and $R_5=2\Omega$. Then Matlab Example 39 shows that the three node voltages are

$v_1 = -2.128\text{V}$, $v_2 = -1.192\text{V}$, and $v_3 = 0.724\text{V}$. The branch currents can then be found by using Ohm's law as follows.

$$i_1 = \frac{v_1}{R_1} = -1.0638\text{A}$$

$$i_2 = \frac{v_1 - v_2}{R_2} = -0.9362\text{A}$$

$$i_3 = \frac{v_2}{R_3} = -0.2979\text{A}$$

$$i_4 = \frac{v_2 - v_3}{R_4} = -0.6383\text{A}$$

$$i_5 = \frac{v_3}{R_5} = 0.3617\text{A}$$

Matlab Example 39

```
>> is = 1;
>> r1 = 2;
>> r2 = 1;
>> r3 = 4;
>> r4 = 3;
>> r5 = 2;
>> g1 = 1/r1;
>> g2 = 1/r2;
>> g3 = 1/r3;
>> g4 = 1/r4;
>> g5 = 1/r5;

>> G = [g1+g2 -g2 0; -g2 g2+g3+g4 -g4; 0 -g4 g4+g5]
G =
    1.5000    -1.0000         0
   -1.0000     1.5833   -0.3333
        0    -0.3333     0.8333

>> i = [-2 0 is]'
i =
    -2
     0
     1

>> v = G\i
v =
   -2.1277
   -1.1915
    0.7234

>> i1 = v(1)/r1
i1 =
   -1.0638
```

Matlab Example 39 (cont.)

```
>> i2 = (v(1)-v(2))/r2
i2 =
   -0.9362

>> i3 = v(2)/r3
i3 =
   -0.2979

>> i4 = (v(2)-v(3))/r4
i4 =
   -0.6383

>> i5 = v(3)/r5
i5 =
    0.3617
```

7.4 Nodal Analysis by Inspection

If you study Example 39 carefully, you will see that the matrix equation given by Eq. (7.31) can be written by inspection without having to go through all of the steps in Eqs. (7.19) – (7.30).

Eq. (7.31) is of the form

$$\mathbf{Gv} = \mathbf{i} \qquad\qquad (7.32)$$

where $\mathbf{G}$ in an $l \times l$ symmetric conductance matrix, $\mathbf{v}$ is an $l \times 1$ vector of node voltages, and $\mathbf{i}$ is a vector of currents representing known currents.

Here are the rules for writing Eq. (7.32) by inspection.

- The matrix $\mathbf{G}$ is symmetric, $g_{kj} = g_{jk}$ and all of the off-diagonal terms are negative or zero.
- The g_{kk} terms are the sum of all conductances connected to node k.
- The g_{kj} terms are the negative sum of the conductances connected to BOTH node k and node j.
- The i_k (the kth component of the vector $\mathbf{i}$) terms are equal to the algebraic sum of the independent currents connected to node k, with *currents entering the node taken as positive*.

We will illustrate these rules in Example 40.

Example 40 – Nodal Analysis by Inspection

Let's solve for the nodal voltages v_1, v_2, and v_3 in the circuit shown in Fig. 7.5 by inspection.

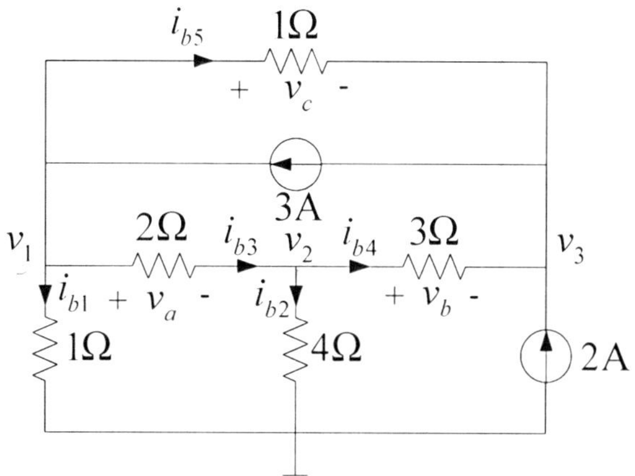

Figure 7.5 Illustrating nodal analysis by inspection

We begin by writing the matrix equation in the form

$$\begin{bmatrix} & & \\ & & \\ & & \end{bmatrix}\begin{bmatrix} v_1 \\ v_2 \\ v_3 \end{bmatrix} = \begin{bmatrix} \\ \\ \end{bmatrix} \tag{7.33}$$

Because we have three nodes, the conductance matrix will be 3 x 3 and the matrix of node voltages will be 3 x 1. We'll begin by filling in the main diagonal of the conductance matrix by just adding the conductances connected to each node. This will give

$$\begin{bmatrix} \frac{1}{1}+\frac{1}{2}+\frac{1}{1} & & \\ & \frac{1}{2}+\frac{1}{3}+\frac{1}{4} & \\ & & \frac{1}{3}+\frac{1}{1} \end{bmatrix}\begin{bmatrix} v_1 \\ v_2 \\ v_3 \end{bmatrix} = \begin{bmatrix} \\ \\ \end{bmatrix} \tag{7.34}$$

Now we will fill in the off-diagonal terms. Remember these will be the negative of the sum of the conductances that are connected to both nodes. For example, a 2Ω resistor is common to nodes 1 and 2 so the g_{12} entry will be $-1/2$. The conductance matrix is symmetric so this will also be the entry for g_{21}. Filling in all the off-diagonal terms will give

$$\begin{bmatrix} \dfrac{1}{1}+\dfrac{1}{2}+\dfrac{1}{1} & -\dfrac{1}{2} & -\dfrac{1}{1} \\[2mm] -\dfrac{1}{2} & \dfrac{1}{2}+\dfrac{1}{3}+\dfrac{1}{4} & -\dfrac{1}{3} \\[2mm] -\dfrac{1}{1} & -\dfrac{1}{3} & \dfrac{1}{3}+\dfrac{1}{1} \end{bmatrix}\begin{bmatrix} v_1 \\ v_2 \\ v_3 \end{bmatrix} = \begin{bmatrix} \\ \\ \end{bmatrix} \qquad (7.35)$$

We now only have to fill in the matrix of known currents on the right-hand side of Eq. (7.35). These will be the sum of the independent current sources *entering* each node. Remember when we developed our equations for nodal analysis we took the sign of currents *leaving* a node as positive. But now we have moved the independent current sources to the right-hand side of the equation so when using the method of inspection we must take currents *entering* a node to be positive. The completed matrix equation will be

$$\begin{bmatrix} \dfrac{1}{1}+\dfrac{1}{2}+\dfrac{1}{1} & -\dfrac{1}{2} & -\dfrac{1}{1} \\[2mm] -\dfrac{1}{2} & \dfrac{1}{2}+\dfrac{1}{3}+\dfrac{1}{4} & -\dfrac{1}{3} \\[2mm] -\dfrac{1}{1} & -\dfrac{1}{3} & \dfrac{1}{3}+\dfrac{1}{1} \end{bmatrix}\begin{bmatrix} v_1 \\ v_2 \\ v_3 \end{bmatrix} = \begin{bmatrix} 3 \\ 0 \\ 2-3 \end{bmatrix} \qquad (7.36)$$

Matlab Example 40 shows the Matlab solution to Eq. (7.36). Note that this is just as easy as writing mesh equations by inspection but in this case we use a conductance matrix instead of a resistance matrix.

The branch currents can then be found by using Ohm's law as follows.

$$i_{b1} = v_1/1\Omega = 1.7368\text{A}$$
$$i_{b2} = v_2/4\Omega = 0.2632\text{A}$$
$$i_{b3} = (v_1 - v_2)/2\Omega = 0.3421\text{A}$$
$$i_{b4} = (v_2 - v_3)/3\Omega = 0.0789\text{A}$$
$$i_{b5} = (v_1 - v_3)/1\Omega = 0.9211\text{A}$$

Other voltages across resistors are

$$v_a = v_1 - v_2 = 0.6842\text{V}$$
$$v_b = v_2 - v_3 = 0.2368\text{V}$$
$$v_c = v_1 - v_3 = 0.9211\text{V}$$

Matlab Example 40

```
>> G = [1+1/2+1 -1/2 -1; -1/2 1/2+1/3+1/4 -1/3; -1 -1/3 1/3+1]
G =
    2.5000   -0.5000   -1.0000
   -0.5000    1.0833   -0.3333
   -1.0000   -0.3333    1.3333

>> i = [3 0 2-3]'
i =
     3
     0
    -1

>> v = G\i
v =
    1.7368
    1.0526
    0.8158

>> ib1 = v(1)/1
ib1 =
    1.7368

>> ib2 = v(2)/4
ib2 =
    0.2632

>> ib3 = (v(1)-v(2))/2
ib3 =
    0.3421

>> ib4 = (v(2)-v(3))/3
ib4 =
    0.0789

>> ib5 = (v(1)-v(3))/1
ib5 =
    0.9211

>> va = v(1)-v(2)
va =
    0.6842

>> vb = v(2)-v(3)
vb =
    0.2368

>> vc = v(1)-v(3)
vc =
    0.9211
```

Example 41 – AC Nodal Analysis by Inspection

In the circuit shown in Fig. 7.6, find the two phasor node voltages V_1 and V_2 and the phasor branch current I_0.

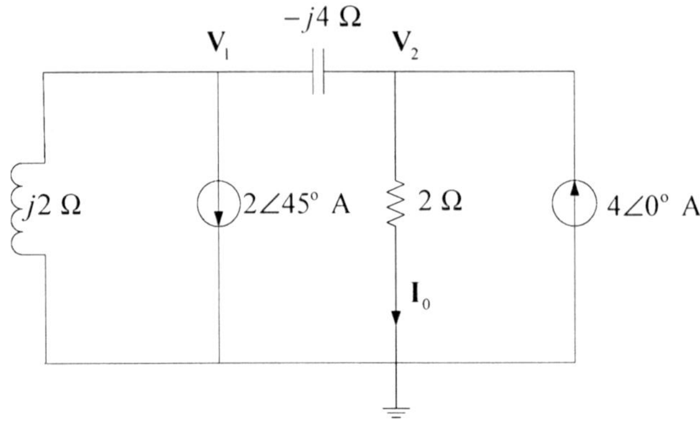

Figure 7.6 Circuit for Example 41

In the case of an AC circuit, the conductances are replaced with admittances. By inspection, we can write the matrix nodal equations as

$$\begin{bmatrix} \dfrac{1}{j2}+\dfrac{1}{-j4} & -\dfrac{1}{-j4} \\[2ex] -\dfrac{1}{-j4} & \dfrac{1}{-j4}+\dfrac{1}{2} \end{bmatrix}\begin{bmatrix} V_1 \\[1ex] V_2 \end{bmatrix}=\begin{bmatrix} -2\angle45° \\[1ex] 4\angle0° \end{bmatrix}$$

which can be rewritten as

$$\begin{bmatrix} \dfrac{-j}{2}+\dfrac{j}{4} & -\dfrac{j}{4} \\[2ex] -\dfrac{j}{4} & \dfrac{j}{4}+\dfrac{1}{2} \end{bmatrix}\begin{bmatrix} V_1 \\[1ex] V_2 \end{bmatrix}=\begin{bmatrix} -2\angle45° \\[1ex] 4\angle0° \end{bmatrix} \tag{7.37}$$

From Matlab Example 41, we see that the solution is

$$\begin{bmatrix} V_1 \\[1ex] V_2 \end{bmatrix}=\begin{bmatrix} 2.0292\angle-125.26° \\[1ex] 7.9137\angle-30.36° \end{bmatrix} \tag{7.38}$$

$$I_0=\frac{V_2}{2}=3.957\angle-30.36° \tag{7.39}$$

Matlab Example 41

```
>> Y = [-j/2+j/4 -j/4; -j/4 j/4+1/2]
Y =
        0 - 0.2500i         0 - 0.2500i
        0 - 0.2500i    0.5000 + 0.2500i

>> I = [-2*exp(j*45*pi/180); 4]
I =
  -1.4142 - 1.4142i
   4.0000

>> V = Y\I
V =
  -1.1716 - 1.6569i
   6.8284 - 4.0000i

>> magV = abs(V)
magV =
    2.0292
    7.9137

>> angV = angle(V)*180/pi
angV =
-125.2644
  -30.3612

>> I0 = V(2)/2
I0 =
   3.4142 - 2.0000i

>> magI0 = abs(I0)
magI0 =
    3.9569

>> angI0 = angle(I0)*180/pi
angI0 =
  -30.3612
```

7.5 Mesh Analysis with Voltage and Current Sources

In Section 7.2, we saw how easy it is to write mesh equations in matrix form by inspection. However, this method works if we have only independent voltage sources. If the circuit contains both independent current sources and independent voltage sources, then we need to take a somewhat different approach. There are a couple of different ways to proceed, but we will use a general method that will allow you to find all currents and voltages in the circuit. We will illustrate the method in Example 42.

Example 42 – Mesh Analysis with Voltage and Current Sources

Consider the circuit shown in Fig. 7.7 that contains both a voltage source and a current source. If we had only voltage sources, then we could solve for the three mesh currents by inspection using the method described in Section 7.2. But now we have a current source in loop 1. However, because this is an independent current source it means that this current source must be the value of the mesh current i_1. Thus, $i_1 = 4$ mA.

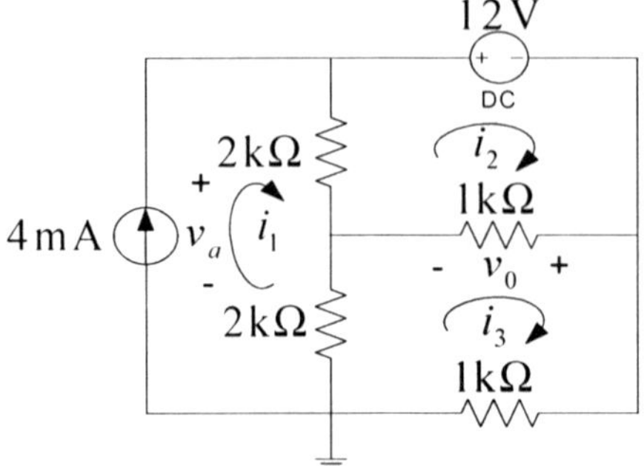

Figure 7.7 Illustrating mesh analysis with both voltage and current sources

Now, in using KVL, we sum the voltages around the loop. But in mesh 1 in Fig. 7.7, we don't know the voltage v_a across the 4mA current source. Let's just call it v_a and treat it as an unknown to be found. Therefore, if we treat v_a as an independent voltage source, we can write the mesh equations for the circuit in Fig. 7.7 by inspection as

$$\begin{bmatrix} 4 & -2 & -2 \\ -2 & 3 & -1 \\ -2 & -1 & 4 \end{bmatrix} \begin{bmatrix} i_1 \\ i_2 \\ i_3 \end{bmatrix} = \begin{bmatrix} v_a \\ -12 \\ 0 \end{bmatrix} \tag{7.40}$$

where we have written the resistances in units of kΩ and the currents will have units of mA. But we have already seen that we know that i_1 is equal to 4mA. Therefore, it is not an unknown, but v_a is. We will therefore rearrange Eq. (7.40) to make v_a, i_2, and i_3 the unknowns. First, we will set $i_1 = 4$ in Eq. (7.40) to get

$$\begin{bmatrix} 4 & -2 & -2 \\ -2 & 3 & -1 \\ -2 & -1 & 4 \end{bmatrix} \begin{bmatrix} 4 \\ i_2 \\ i_3 \end{bmatrix} = \begin{bmatrix} v_a \\ -12 \\ 0 \end{bmatrix} \tag{7.41}$$

We will now expand Eq. (7.41) to get the three equations

$$16 - 2i_2 - 2i_3 = v_a \tag{7.42}$$
$$-8 + 3i_2 - i_3 = -12 \tag{7.43}$$
$$-8 - i_2 + 4i_3 = 0 \tag{7.44}$$

Rearranging Eqs. (7.42) – (7.44) to put the unknowns v_a, i_2, and i_3 on the left-hand side and the known values on the right-hand side we obtain

$$-v_a - 2i_2 - 2i_3 = -16 \tag{7.45}$$

$$0v_a + 3i_2 - i_3 = -4 \tag{7.46}$$

$$0v_a - i_2 + 4i_3 = 8 \tag{7.47}$$

Rewriting Eqs. (7.45) – (7.47) in matrix form we obtain

$$\begin{bmatrix} -1 & -2 & -2 \\ 0 & 3 & -1 \\ 0 & -1 & 4 \end{bmatrix} \begin{bmatrix} v_a \\ i_2 \\ i_3 \end{bmatrix} = \begin{bmatrix} -16 \\ -4 \\ 8 \end{bmatrix} \tag{7.48}$$

This equation is of the form

$$\mathbf{Mu = k} \tag{7.49}$$

where $\mathbf{M}$ is no longer a symmetric matrix, $\mathbf{u}$ is a vector of the unknowns v_a, i_2, and i_3, and $\mathbf{k}$ is a vector of known values. Matlab Example 42 gives the solution of Eq. (7.48).

From this solution we see that $v_a = u(1) = 13.8182$ V. The unknown mesh currents are $i_2 = u(2) = -0.7273$ mA and $i_3 = u(3) = 1.8182$ mA. From these mesh currents, we can find all other voltages and branch currents in the circuit. For example, the voltage v_0 shown in Fig. 7.7 is given by

$$v_0 = 1 \, k\Omega \left(i_2 - i_3 \right) = -2.5455 \text{ V} .$$

Matlab Example 42

```
>> M = [-1 -2 -2; 0 3 -1; 0 -1 4]
M =
      -1      -2      -2
       0       3      -1
       0      -1       4

>> k = [-16 -4 8]'
k =
     -16
      -4
       8

>> u = M\k
u =
     13.8182
     -0.7273
      1.8182
```

Matlab Example 42 (cont.)

```
>> i1 = 4
i1 =
      4

>> i2 = u(2)
i2 =
    -0.7273

>> i3 = u(3)
i3 =
     1.8182

>> va = u(1)
va =
    13.8182

>> v0 = 1*(i2-i3)
v0 =
    -2.5455
```

Example 43 – AC Mesh with Voltage and Current Sources

In the circuit shown in Fig. 7.8, find the two phasor mesh currents I_1 and I_2 and the phasor branch current I_0.

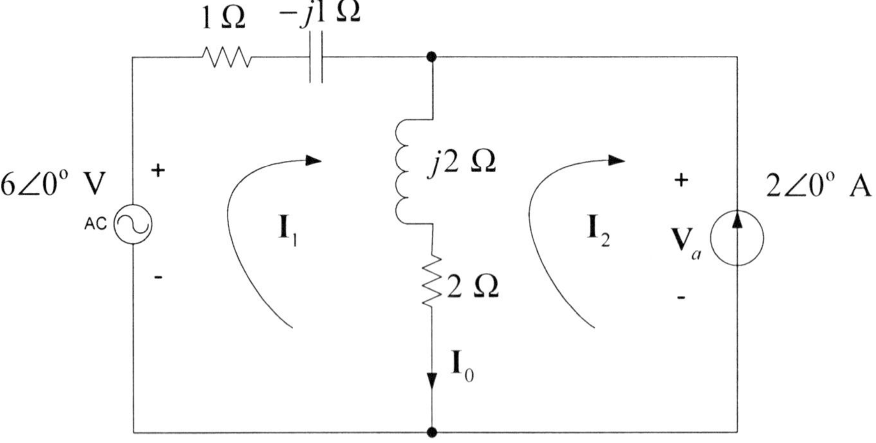

Figure 7.8 Circuit for Example 43

We assume we know the voltage V_a across the current source and then write the matrix mesh equations by inspection as follows:

$$\begin{bmatrix} 3+j1 & -2-j2 \\ -2-j2 & 2+j2 \end{bmatrix}\begin{bmatrix} \mathbf{I}_1 \\ \mathbf{I}_2 \end{bmatrix} = \begin{bmatrix} 6 \\ -\mathbf{V}_a \end{bmatrix} \tag{7.50}$$

We know that the mesh current $\mathbf{I}_2 = -2\angle 0^\circ$ A. Therefore, we can rewrite Eq. (7.50) as

$$\begin{bmatrix} 3+j1 & -2-j2 \\ -2-j2 & 2+j2 \end{bmatrix}\begin{bmatrix} \mathbf{I}_1 \\ -2 \end{bmatrix} = \begin{bmatrix} 6 \\ -\mathbf{V}_a \end{bmatrix} \tag{7.51}$$

We can expand Eq. (7.51) to get the two equations

$$\left(3+j1\right)\mathbf{I}_1 + 4 + j4 = 6 \tag{7.52}$$
$$\left(-2-j2\right)\mathbf{I}_1 - 4 - j4 = -\mathbf{V}_a \tag{7.53}$$

Rearranging Eqs. (7.52) – (7.53) to put the unknowns I_1 and V_a on the left-hand side and the known values on the right-hand side, we obtain

$$\left(3+j1\right)\mathbf{I}_1 + 0\mathbf{V}_a = 2 - j4 \tag{7.54}$$
$$\left(-2-j2\right)\mathbf{I}_1 + 1\mathbf{V}_a = 4 + j4 \tag{7.55}$$

Rewriting Eqs. (7.54) – (7.55) in matrix form, we obtain

$$\begin{bmatrix} 3+j1 & 0 \\ -2-j2 & 1 \end{bmatrix}\begin{bmatrix} \mathbf{I}_1 \\ \mathbf{V}_a \end{bmatrix} = \begin{bmatrix} 2-j4 \\ 4+j4 \end{bmatrix} \tag{7.56}$$

From Matlab Example 43, we see that the solution of Eq. (7.56) is

$$U = \begin{bmatrix} \mathbf{I}_1 \\ \mathbf{V}_a \end{bmatrix} = \begin{bmatrix} 1.414\angle -81.87^\circ \\ 7.376\angle 12.53^\circ \end{bmatrix} \tag{7.57}$$

$$\mathbf{I}_0 = \mathbf{I}_1 - \mathbf{I}_2 = 2.61\angle -32.47^\circ \text{ A} \tag{7.58}$$

From this solution, we see that $\mathbf{V}_a = u(1) = 7.376\angle 12.53^\circ$ V. The unknown mesh current is $\mathbf{I}_1 = u(1) = 1.414\angle -81.87^\circ$ A. From these mesh currents, we can find all other voltages and branch currents in the circuit. Note that this is the same circuit that we solved in Example 28 using source transformation where we found the same value for $\mathbf{I}_0$ as in Eq. (7.58).

Matlab Example 43

```
>> M = [3+1j 0; -2-2j 1]
M =
   3.0000 + 1.0000i          0
  -2.0000 - 2.0000i     1.0000

>> k = [2-4j; 4+4j]
k =
   2.0000 - 4.0000i
   4.0000 + 4.0000i

>> u = M\k
u =
   0.2000 - 1.4000i
   7.2000 + 1.6000i

>> magu = abs(u)
magu =
    1.4142
    7.3756

>> angu = angle(u)*180/pi
angu =
  -81.8699
   12.5288

>> I2 = -2
I2 =
    -2

>> I1 = u(1)
I1 =
   0.2000 - 1.4000i

>> I0 = I1 - I2
I0 =
   2.2000 - 1.4000i

>> magI0 = abs(I0)
magI0 =
    2.6077

>> angI0 = angle(I0)*180/pi
angI0 =
  -32.4712
```

7.6 Nodal Analysis with Voltage and Current Sources

In Section 7.4, we saw how easy it is to write nodal equations in matrix form by inspection. However, this method works if we have only independent current sources. If the circuit contains both independent voltage sources and independent current sources, then we need to take a somewhat different approach. There are a couple of different ways to proceed, but we will use a general method that will allow you to find all currents and voltages in the circuit. We will illustrate the method in Example 44.

Example 44 – Nodal Analysis with Voltage and Current Sources

Consider the circuit shown in Fig. 7.9 that contains both a voltage source and two current sources. If we had only current sources, then we could solve for the three node voltages by inspection using the method described in Section 7.4. But now we have a voltage between nodes 1 and 3. However, because this is an independent voltage source it means that this voltage source must be equal to $v_1 - v_3$. Thus, $v_3 = v_1 - 3V$ and we really have only two unknown node voltages, v_1 and v_2.

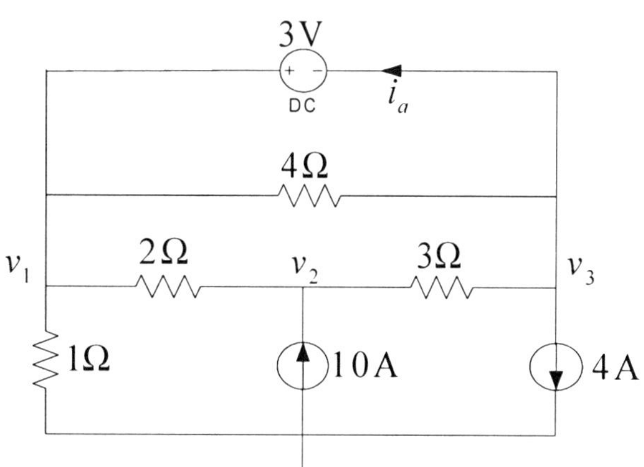

Figure 7.9 Illustrating nodal analysis with both voltage and current sources

Now, in using KCL, we sum the currents leaving a node. But, in Fig. 7.9, we don't know the current i_a through the 3V voltage source, which leaves node 3 and enters node 1. Let's just call it i_a and treat it as an unknown to be found. Therefore, if we treat i_a as an independent current source, we can write the nodal equations for the circuit in Fig. 7.9 by inspection as

$$
\begin{bmatrix}
\dfrac{1}{4}+\dfrac{1}{2}+\dfrac{1}{1} & -\dfrac{1}{2} & -\dfrac{1}{4} \\[2mm]
-\dfrac{1}{2} & \dfrac{1}{2}+\dfrac{1}{3} & -\dfrac{1}{3} \\[2mm]
-\dfrac{1}{4} & -\dfrac{1}{3} & \dfrac{1}{4}+\dfrac{1}{3}
\end{bmatrix}
\begin{bmatrix} v_1 \\ v_2 \\ v_3 \end{bmatrix}
=
\begin{bmatrix} i_a \\ 10 \\ -4-i_a \end{bmatrix}
\qquad (7.59)
$$

where we have written the conductances using the normal rules for nodal analysis by inspection from Section 7.4. But we have already seen that we know that $v_3 = v_1 - 3\text{V}$. Therefore, it is not an unknown, but i_a is. We will therefore rearrange Eq. (7.59) to make v_1, v_2, and i_a the unknowns. First, we will set $v_3 = v_1 - 3$ in Eq. (7.59) to get

$$\begin{bmatrix} \frac{1}{4}+\frac{1}{2}+\frac{1}{1} & -\frac{1}{2} & -\frac{1}{4} \\ -\frac{1}{2} & \frac{1}{2}+\frac{1}{3} & -\frac{1}{3} \\ -\frac{1}{4} & -\frac{1}{3} & \frac{1}{4}+\frac{1}{3} \end{bmatrix} \begin{bmatrix} v_1 \\ v_2 \\ v_1 - 3 \end{bmatrix} = \begin{bmatrix} i_a \\ 10 \\ -4 - i_a \end{bmatrix} \tag{7.60}$$

We will now expand Eq. (7.60) to get the three equations

$$\left(\frac{1}{4}+\frac{1}{2}+\frac{1}{1}\right)v_1 - \frac{1}{2}v_2 - \frac{1}{4}(v_1 - 3) = i_a \tag{7.61}$$

$$-\frac{1}{2}v_1 + \left(\frac{1}{2}+\frac{1}{3}\right)v_2 - \frac{1}{3}(v_1 - 3) = 10 \tag{7.62}$$

$$-\frac{1}{4}v_1 - \frac{1}{3}v_2 + \left(\frac{1}{4}+\frac{1}{3}\right)(v_1 - 3) = -4 - i_a \tag{7.63}$$

Rearranging Eqs. (7.61) – (7.63) to put the unknowns v_1, v_2, and i_a on the left-hand side and the known values on the right-hand side, we obtain

$$\left(\frac{1}{2}+\frac{1}{1}\right)v_1 - \frac{1}{2}v_2 - 1i_a = -\frac{3}{4} \tag{7.64}$$

$$\left(-\frac{1}{2}-\frac{1}{3}\right)v_1 + \left(\frac{1}{2}+\frac{1}{3}\right)v_2 + 0i_a = 10 - 1 \tag{7.65}$$

$$\frac{1}{3}v_1 - \frac{1}{3}v_2 + 1i_a = -4 + 3\left(\frac{1}{4}+\frac{1}{3}\right) \tag{7.66}$$

Rewriting Eqs. (7.64) – (7.66) in matrix form, we obtain

$$\begin{bmatrix} \left(\frac{1}{2}+\frac{1}{1}\right) & -\frac{1}{2} & -1 \\ \left(-\frac{1}{2}-\frac{1}{3}\right) & \left(\frac{1}{2}+\frac{1}{3}\right) & 0 \\ \frac{1}{3} & -\frac{1}{3} & 1 \end{bmatrix} \begin{bmatrix} v_1 \\ v_2 \\ i_a \end{bmatrix} = \begin{bmatrix} -\frac{3}{4} \\ 9 \\ -4+\frac{3}{4}+1 \end{bmatrix} \tag{7.67}$$

This equation is of the form

$$\mathbf{M u} = \mathbf{k} \qquad (7.68)$$

where $\mathbf{M}$ is no longer a symmetric matrix, $\mathbf{u}$ is a vector of the unknowns v_1, v_2, and i_a, and $\mathbf{k}$ is a vector of known values. Matlab Example 44 gives the solution of Eq. 7.67.

Matlab Example 44

```
>> M = [1/2+1 -1/2 -1; -1/2-1/3 1/2+1/3 0; 1/3 -1/3 1]
M =
    1.5000    -0.5000    -1.0000
   -0.8333     0.8333          0
    0.3333    -0.3333     1.0000

>> k = [-3/4 9 -4+3/4+1]'
k =
   -0.7500
    9.0000
   -2.2500

>> u = M\k
u =
    6.0000
   16.8000
    1.3500

>> v1 = u(1)
v1 =
     6

>> v2 = u(2)
v2 =
   16.8000

>> v3 = v1 - 3
v3 =
     3

>> ia = u(3)
ia =
    1.3500

>> i23 = (v2-v3)/3
i23 =
    4.6000

>> i21 = (v2-v1)/2
i21 =
    5.4000
```

From this solution we see that $i_a = u(3) = 1.35$ A.   The node voltages are $v_1 = u(1) = 6\text{V}$, $v_2 = u(2) = 16.8\text{V}$, and $v_3 = v_1 - 3\text{V} = 3\text{V}$.   From these node voltages, we can find all other branch currents in the circuit. For example, the voltage i_{23} through the 3Ω resistor is given by $i_{23} = (v_2 - v_3)/3\Omega = 4.6\text{A}$. Similarly, the voltage i_{21} through the 2Ω resistor is given by $i_{21} = (v_2 - v_1)/2\Omega = 5.4\text{A}$. Note that these two currents add up to the 10A current source.

Example 45 – AC Nodal with Voltage and Current Sources

Use nodal analysis to find the phasor current $\mathbf{I}_0$ in the circuit shown in Fig. 7.10.

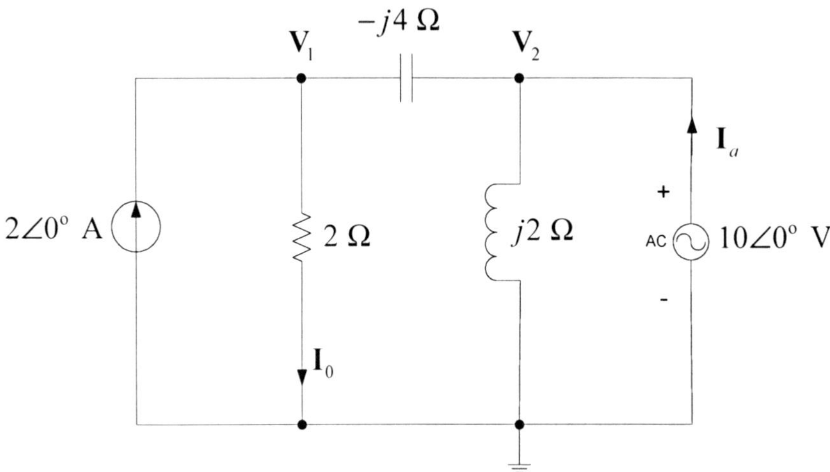

Figure 7.10 Circuit for Example 45

We will begin by using nodal analysis to solve for the phasor voltages V1 and V2. The current through the 2Ω resistor will then be $\mathbf{I}_0 = \mathbf{V}_1/2\ \Omega$. If we assume we know the current $\mathbf{I}_a$ through the 10 V voltage source, we can write the nodal matrix equation by inspection as

$$\begin{bmatrix} \dfrac{1}{2} + \dfrac{j}{4} & -\dfrac{j}{4} \\ -\dfrac{j}{4} & \dfrac{j}{4} - \dfrac{j}{2} \end{bmatrix} \begin{bmatrix} \mathbf{V}_1 \\ \mathbf{V}_2 \end{bmatrix} = \begin{bmatrix} 2\angle 0^\circ \\ \mathbf{I}_a \end{bmatrix} \tag{7.69}$$

We know that the node voltage $\mathbf{V}_2 = 10\angle 0^\circ$ V. Therefore, we can rewrite Eq. (7.69) as

$$\begin{bmatrix} \dfrac{1}{2}+\dfrac{j}{4} & -\dfrac{j}{4} \\ -\dfrac{j}{4} & \dfrac{j}{4}-\dfrac{j}{2} \end{bmatrix}\begin{bmatrix} \mathbf{V}_1 \\ 10 \end{bmatrix} = \begin{bmatrix} 2\angle 0^\circ \\ \mathbf{I}_a \end{bmatrix} \tag{7.70}$$

We can expand Eq. (7.70) to get the two equations

$$\left(\dfrac{1}{2}+\dfrac{j}{4}\right)\mathbf{V}_1 - \dfrac{j10}{4} = 2 \tag{7.71}$$

$$\left(-\dfrac{j}{4}\right)\mathbf{V}_1 + 10\left(\dfrac{j}{4}-\dfrac{j}{2}\right) = \mathbf{I}_a \tag{7.72}$$

Rearranging Eqs. (7.71) – (7.72) to put the unknowns $\mathbf{V}_1$ and $\mathbf{I}_a$ on the left-hand side and the known values on the right-hand side, we obtain

$$\left(\dfrac{1}{2}+\dfrac{j}{4}\right)\mathbf{V}_1 + 0\mathbf{I}_a = 2 + \dfrac{j10}{4} \tag{7.73}$$

$$\left(-\dfrac{j}{4}\right)\mathbf{V}_1 - 1\mathbf{I}_a = -10\left(\dfrac{j}{4}-\dfrac{j}{2}\right) \tag{7.74}$$

Rewriting Eqs. (7.73) – (7.745) in matrix form we obtain

$$\begin{bmatrix} \dfrac{1}{2}+\dfrac{j}{4} & 0 \\ -\dfrac{j}{4} & -1 \end{bmatrix}\begin{bmatrix} \mathbf{V}_1 \\ \mathbf{I}_a \end{bmatrix} = \begin{bmatrix} 2 + \dfrac{j10}{4} \\ -10\left(\dfrac{j}{4}-\dfrac{j}{2}\right) \end{bmatrix} \tag{7.75}$$

From Matlab Example 45, we see that the solution of Eq. (7.75) is

$$\mathbf{U} = \begin{bmatrix} \mathbf{V}_1 \\ \mathbf{I}_a \end{bmatrix} = \begin{bmatrix} 5.727\angle 24.76^\circ \\ 3.847\angle -81.03^\circ \end{bmatrix} \tag{7.76}$$

$$\mathbf{I}_0 = \dfrac{\mathbf{V}_1}{2} = 2.864\angle 24.76^\circ \text{ A} \tag{7.77}$$

Matlab Example 45

```
>> M = [1/2+j/4 0; -j/4 -1]
M =
   0.5000 + 0.2500i          0
        0 - 0.2500i   -1.0000

>> k = [2+j*10/4; -10*(j/4-j/2)]
k =
   2.0000 + 2.5000i
        0 + 2.5000i

>> u = M\k
u =
   5.2000 + 2.4000i
   0.6000 - 3.8000i

>> magu = abs(u)
magu =
    5.7271
    3.8471

>> angu = angle(u)*180/pi
angu =
   24.7751
  -81.0274

>> I0 = u(1)/2
I0 =
   2.6000 + 1.2000i

>> magI0 = abs(I0)
magI0 =
    2.8636

>> angI0 = angle(I0)*180/pi
angI0 =
   24.7751
```

Problems

7.1   In the circuit shown below, if $r1 = 6\Omega$, $r2 = 3\Omega$, and $r3 = 4\Omega$, use mesh analysis to find the voltage $v1$ and the three currents $i1$, $i2$, and $i3$, and compare your results to those of Example 8 in Chapter 3.

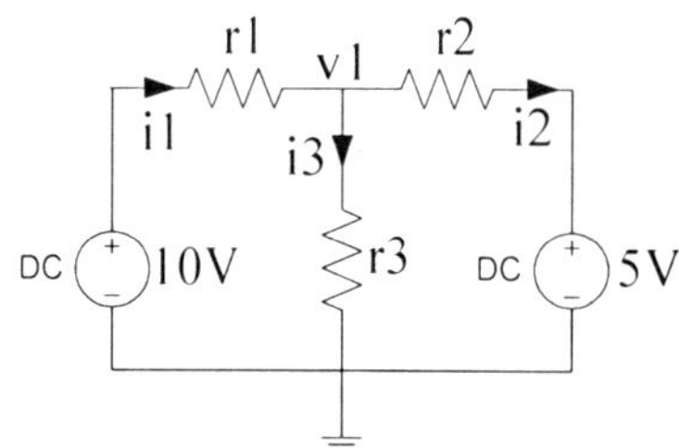

7.2   In the circuit below, if $r1 = 4\ k\Omega$, $r2 = 8\ k\Omega$, and $r3 = 12\ k\Omega$, use mesh analysis to find the voltage $v1$ and the three currents, $i1$, $i2$, and $i3$.

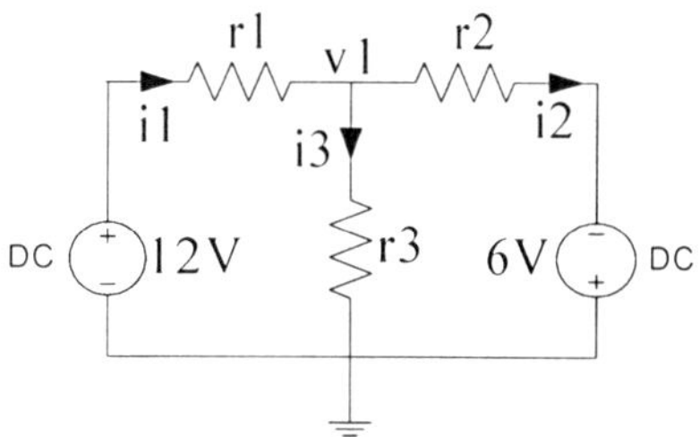

7.3   Repeat Problem 7.2 for values of $r1 = 1\ k\Omega$, $r2 = 1\ k\Omega$, and $r3 = 2\ k\Omega$.

Chapter 8

Circuit Theorems

In Chapter 7, we saw that mesh and nodal analysis could be used to find all of the voltages and currents in a circuit. If we are interested in finding only a single voltage or current, then sometimes we can use some circuit theorem to our advantage. In this chapter, we will describe how to use the method of superposition and also how to find the voltage across or current through a particular element using Thevenin's theorem or Norton's theorem. Finally, we will show the conditions under which the maximum power can be delivered to a load.

8.1 Superposition

A *linear circuit* is one whose output is directly proportional to its input. Linear circuits obey both the properties of *homogeneity* (scaling) and *additivity*. Because the circuit is linear, we can find the response of the circuit to each source acting alone, and then add them up to find the response of the circuit to all sources acting together. This is known as the *superposition principle*. The *superposition principle* states that the voltage across (or the current through) an element in a linear circuit is the algebraic sum of the voltages across (or currents through) that element due to each independent source acting alone.

To have an independent source act alone, we must turn off all other independent sources in the circuit, which means we must set all other independent sources to zero. Setting an independent voltage source to zero is equivalent to making it a *short circuit* since the voltage across a short circuit must be zero. Setting an independent current source to zero is equivalent to making it an *open circuit* since the current through an open circuit must be zero.

The following steps are used in applying the superposition principle.

1. Turn off all independent sources except one. Find the output (voltage or current) due to the active source.
2. Repeat step 1 for each of the other independent sources.
3. Find the total output by adding algebraically all of the results found in steps 1 & 2 above.

In some cases, but certainly not all, superposition can simplify the analysis.

Example 46 – Superposition

In the circuit shown in Fig. 8.1, find the voltage v_0 using superposition and compare your result with that given in Example 42 in Chapter 7.

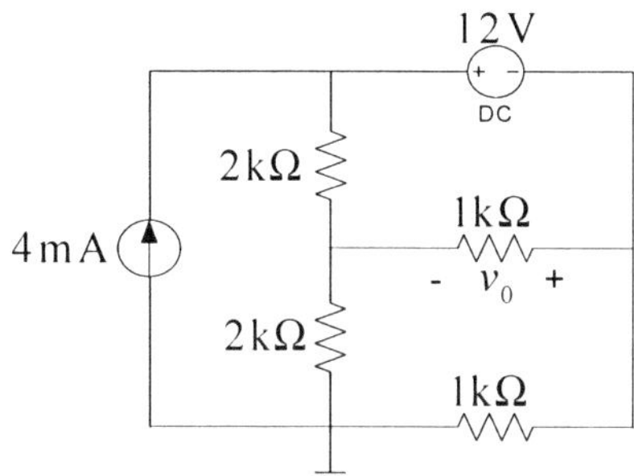

Figure 8.1 Example 46

We will find the value of v_0 due only to the current source and then find the value of v_0 due only to the voltage source and add the results to get the actual value of v_0. Fig. 8.2a shows the circuit with the voltage source set equal to zero (short circuit). This circuit is redrawn in Fig. 8.2b to show that v_0 is measured across the 1 kΩ and 2 kΩ resistors in parallel. This combination is in series with another 2 kΩ resistor and this series combination is then in parallel with the 1 kΩ resistor.

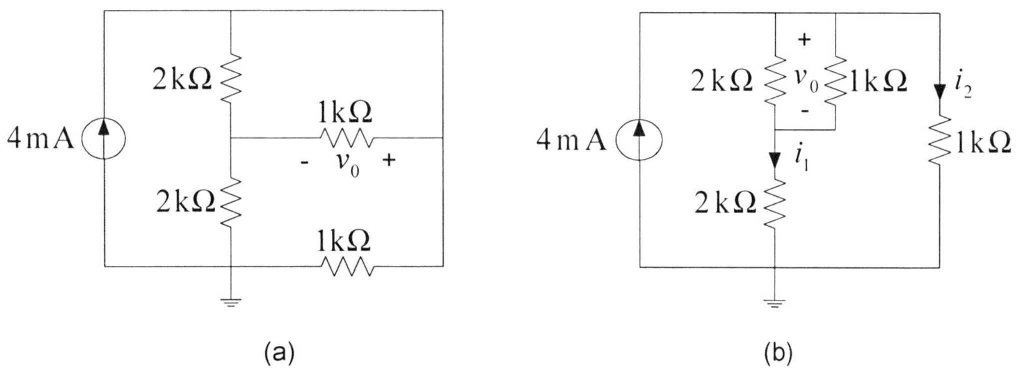

Figure 8.2 Example 46 – Step 1

We can first use current division in Fig. 8.2b to find i_1.

$$i_1 = \frac{1\ k\Omega}{1\ k\Omega + 2\ k\Omega + \dfrac{1\ k\Omega \times 2\ k\Omega}{1\ k\Omega + 2\ k\Omega}}\, 4\ mA = 1.0909\ mA \quad (8.1)$$

The voltage v_0 due only to the current source is then given by

$$v_{01} = i_1 \frac{1 \text{ k}\Omega \times 2 \text{ k}\Omega}{1 \text{ k}\Omega + 2 \text{ k}\Omega} = 0.7273 \text{ V} \qquad (8.2)$$

Fig. 8.3a shows the circuit with the current source set equal to zero (open circuit). This circuit is redrawn in Fig. 8.3b to show that v_0 is measured across the 1kΩ resistor in parallel with a pair of 1kΩ and 2kΩ resistors in series. This combination is in series with another 2kΩ resistor. The whole circuit is in the form of a voltage divider. We can therefore write the voltage v_0 due only to the voltage source as

$$v_{02} = \frac{\dfrac{1 \text{ k}\Omega \times 3 \text{ k}\Omega}{1 \text{ k}\Omega + 3 \text{ k}\Omega}}{\dfrac{1 \text{ k}\Omega \times 3 \text{ k}\Omega}{1 \text{ k}\Omega + 3 \text{ k}\Omega} + 2 \text{ k}\Omega}(-12 \text{ V}) = -3.2727 \text{ V} \qquad (8.3)$$

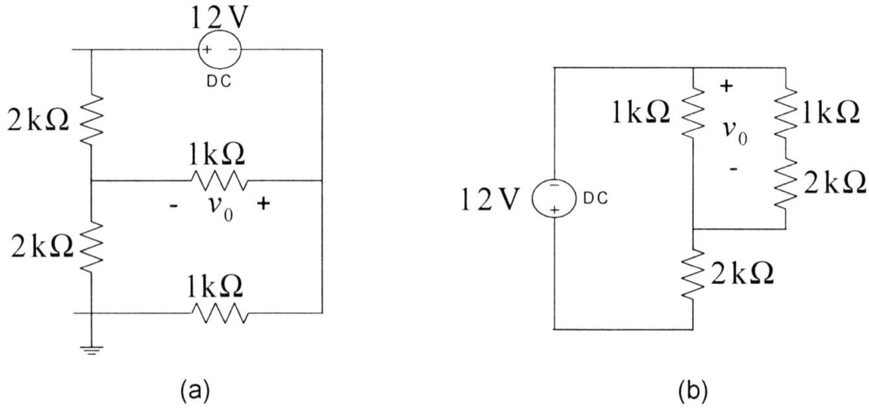

Figure 8.3 Example 46 – Step 2

The voltage v_0 in Fig. 8.1 is then the sum of Eqs. (8.2) and (8.3). Thus,

$$v_0 = v_{01} + v_{02} = 0.7273 \text{ V} - 3.2727 \text{ V} = -2.5455 \text{ V} \qquad (8.4)$$

which agrees with the result in Example 42 in Chapter 7. The calculations for this example are shown in Matlab Example 46.

Matlab Example 46

```
>> i1 = (1/(1+2+2/3))*4
i1 =
    1.0909

>> v01 = i1*2/3
v01 =
    0.7273

>> v02 = (0.75/2.75)*(-12)
v02 =
   -3.2727

>> v0 = v01 + v02
v0 =
   -2.5455
```

Example 47 – AC Superposition

In the circuit shown in Fig. 8.4, find the voltage $v_0(t)$ and plot the result using Matlab. Note that, in this circuit, the two voltage sources have different frequencies. This means that we *must* use superposition to solve this problem. We can't use mesh analysis, for example, because the impedances of the inductor and capacitor will be different for the two different voltage sources.

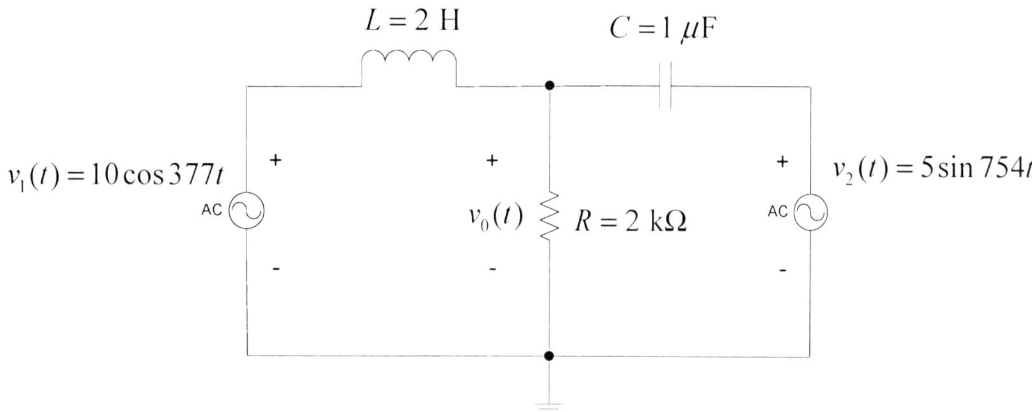

Figure 8.4 Circuit for Example 47

In step 1, we will turn off $v_2(t)$ and consider only the voltage source $v_1(t)$ as shown in Fig. 8.5. The impedance $\mathbf{Z}_1$ of the resistor and capacitor in parallel in Fig. 8.5 can be written as

$$\mathbf{Z}_1 = \frac{R\left(-j/\omega_1 C\right)}{R - \dfrac{j}{\omega_1 C}} = \frac{R}{1 + j\omega_1 RC} \tag{8.5}$$

We can then use voltage division to write the phasor voltage $\mathbf{V}_{01}$ in Fig. 8.5 as

$$\mathbf{V}_{01} = 10\angle 0^\circ \frac{\mathbf{Z}_1}{\mathbf{Z}_1 + j\omega_1 L} = V_{01m}\angle\phi_{01} \tag{8.6}$$

The resulting contribution of voltage source 1 to $v_0(t)$ will be

$$v_{01}(t) = V_{01m}\cos\left(\omega_1 t + \phi_{01}\right) \tag{8.7}$$

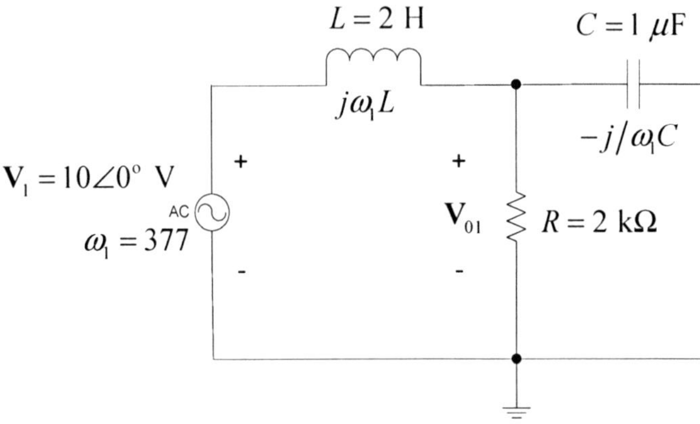

Figure 8.5 Example 47 – Step 1

In step 2, we will turn off $v_1(t)$ and consider only the voltage source $v_2(t)$ as shown in Fig. 8.6. The impedance $\mathbf{Z}_2$ of the resistor and inductor in parallel in Fig. 8.6 can be written as

$$\mathbf{Z}_2 = \frac{R\left(j\omega_2 L\right)}{R + j\omega_2 L} \tag{8.8}$$

We can then use voltage division to write the phasor voltage $\mathbf{V}_{02}$ in Fig. 8.6 as

$$\mathbf{V}_{02} = 5\angle -90^\circ \frac{\mathbf{Z}_2}{\mathbf{Z}_2 - \dfrac{j}{\omega_2 C}} = V_{02m}\angle\phi_{02} \tag{8.9}$$

The resulting contribution of voltage source 2 to $v_0(t)$ will be

$$v_{02}(t) = V_{02m}\cos\left(\omega_2 t + \phi_{02}\right) \tag{8.10}$$

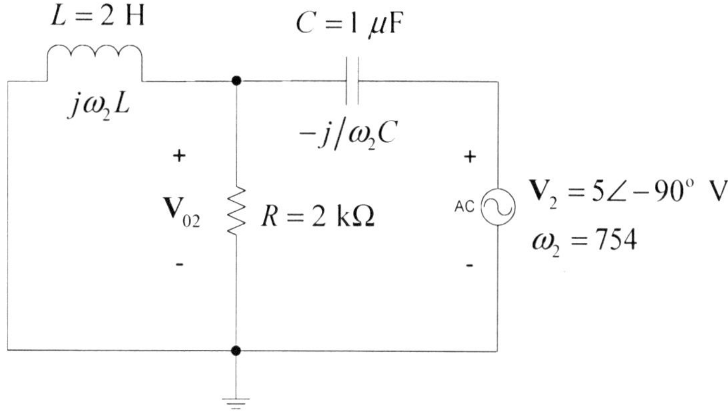

Figure 8.6 Example 47 – Step 2

The total voltage $v_0(t)$ is then found by adding Eqs. (8.7) and (8.10) to obtain

$$v_0(t) = v_{01}(t) + v_{02}(t) \tag{8.11}$$

Matlab Example 47 uses Eqs. (8.5) – (8.11) to obtain the plot shown in Fig. 8.7. Note that the output voltage $v_0(t)$ is no longer a sinusoidal signal but is the sum of two sinusoidal signals of different frequencies.

Matlab Example 47

```
>> C = 10^-6;
>> R = 2000;
>> L = 2;
>> w1 = 377;
>> w2 = 754;
>> Vm1 = 10;
>> Vm2 = 5;
>> Z1 = R/(1+j*w1*R*C)
Z1 =
   1.2751e+003 -9.6142e+002i

>> V01 = Vm1*Z1/(Z1+j*w1*L)
V01 =
   10.9371 - 5.7609i

>> V01m = abs(V01)
V01m =
    12.3616
```

Matlab Example 47 (cont.)

```
>> phi01 = angle(V01)
phi01 =
   -0.4848

>> Z2 = j*R*w2*L/(R+j*w2*L)
Z2 =
  7.2491e+002 +9.6142e+002i

>> V02 = -j*Vm2*Z2/(Z2-j/(w2*C))
V02 =
   7.2989 - 1.3265i

>> V02m = abs(V02)
V02m =
    7.4185

>> phi02 = angle(V02)
phi02 =
   -0.1798

>> t = linspace(0,4*pi/w1,100);
>> v01 = V01m.*cos(w1.*t+phi01);
>> v02 = V02m.*cos(w2.*t+phi02);
>> v0 = v01 + v02;
>> curves = [v01;v02;v0];
>> plot(t,curves)
```

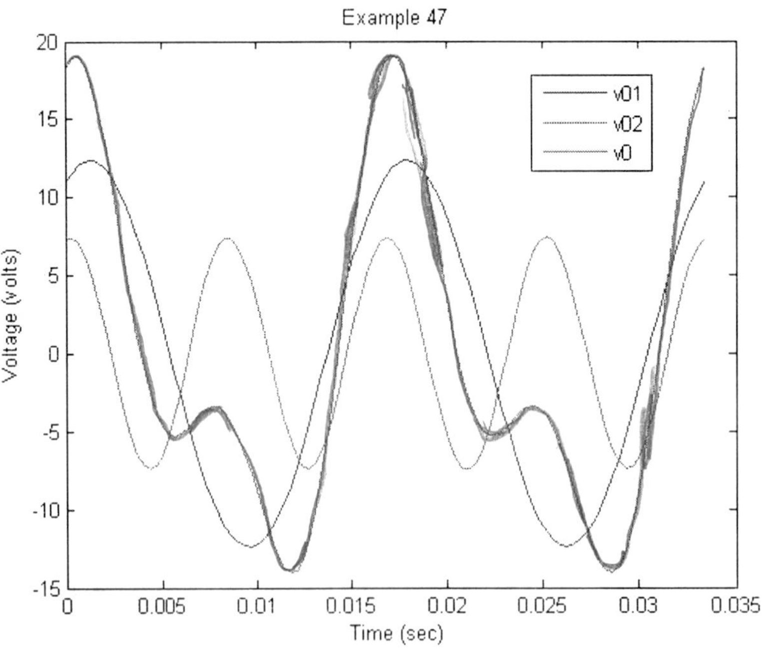

Figure 8.7 Plot resulting from Matlab Example 47

8.2 Thevenin's and Norton's Theorems

In many applications, we want to find the response to a particular element which may, at least at the design stage, be variable as shown in Fig. 8.8. Each time the variable element changes, we have to re-analyze the entire circuit. To avoid this, we would like to have a technique that replaces the linear circuit by something simple that facilitates the analysis. A good approach would be to have a simple *equivalent circuit* to replace everything in the circuit except for the variable part (the load).

What measurements can we make on the linear circuit that will characterize its behavior? Two things we can measure are its open circuit voltage and short circuit current as shown in Fig. 8.9.

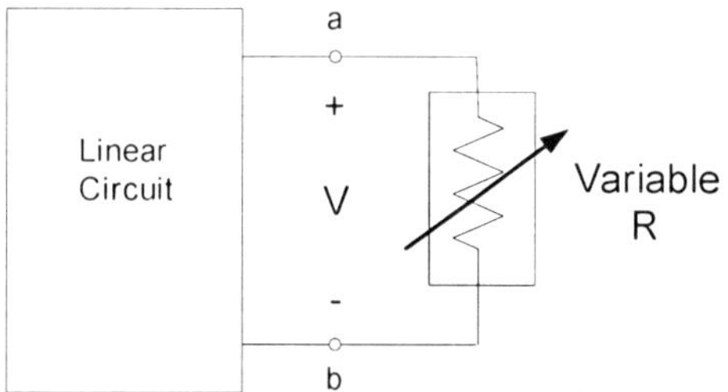

Figure 8.8 Measuring the response of a variable load

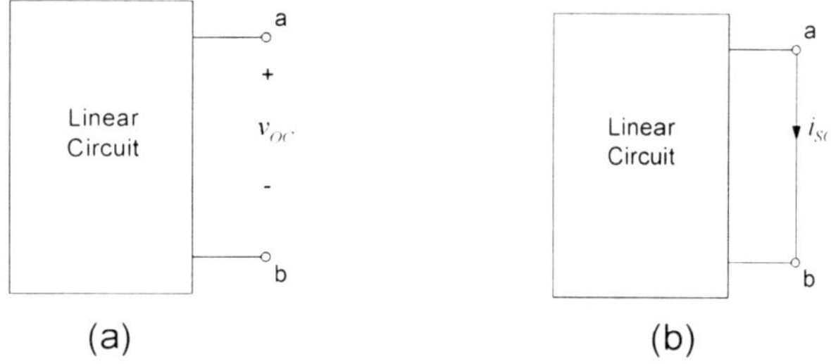

Figure 8.9 Measuring the open circuit voltage and the short circuit current

Given the two measurements in Fig. 8.9 can we find a simple circuit that is equivalent to the linear circuit shown in Fig. 8.9? In fact, the two circuits shown in Fig. 8.10 will work. Note that each circuit gives the correct values for the open circuit voltage and the short circuit current.

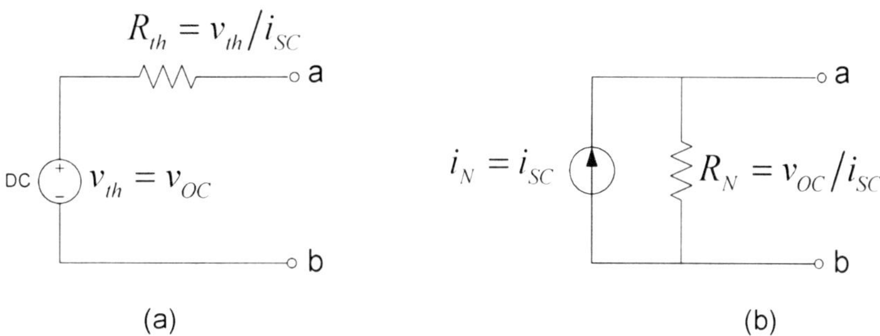

Figure 8.10 a) Thevenin equivalent circuit
b) Norton equivalent circuit

The Thevenin resistance R_{th} and the Norton resistance R_N are equal and given by the open circuit voltage divided by the short circuit current. Note that, in the above, we have chosen the direction of i_{SC} such that it flows out of the terminal which is taken as the positive side of v_{OC}. This is important, since otherwise we would have to introduce a minus sign in the definition of R_{Th}.

This Thevenin or Norton resistance can also be determined by finding the equivalent resistance when all independent sources are turned off (voltage sources become short circuits and current sources become open circuits).

Thevenin's theorem states that a linear two-terminal resistive circuit can be replaced by an equivalent circuit consisting of a voltage source v_{th} in series with a resistor R_{th}, where v_{th} is the open-circuit voltage at the terminals, and R_{th} is the input or equivalent resistance at the terminals when the independent sources are all turned off.

Norton's theorem states that a linear two terminal resistive circuit can be replaced by an equivalent circuit consisting of a current source i_N in parallel with a resistor R_N, where i_N is the short-circuit current through the terminals, and R_N is the input or equivalent resistance at the terminals when all independent sources are all turned off.

Example 48 – Thevenin's Theorem

In this example, we will use Thevenin's theorem to find the voltage v_0 in the circuit shown in Fig. 8.11. This is the same problem we did in Example 46 using superposition. In using Thevenin's theorem to solve for a voltage (or current), the first step is to remove the load resistor across which (or through which) you want to measure the voltage (or current). Thus, if we remove the 1 kΩ load resistor in Fig. 8.11 we get the circuit shown in Fig. 8.12.

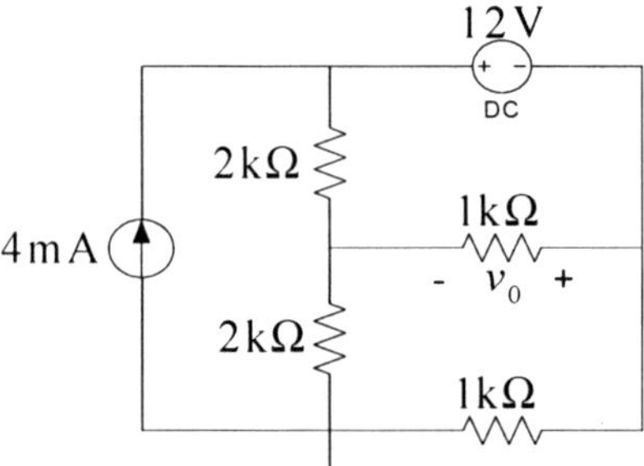

Figure 8.11 Example 48

To find the open circuit voltage v_{OC}, we need to find the voltages v_b and v_c. We can find these if we know the loop current i_2 because we know the mesh current i_1 is 4 mA. Therefore, let's write the mesh equations by inspection.

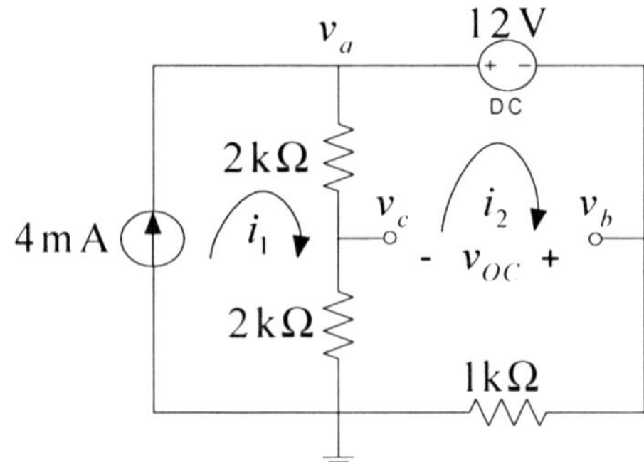

Figure 8.12 Example 48 – Finding v_{OC}

$$\begin{bmatrix} 4k\Omega & -4k\Omega \\ -4k\Omega & 5k\Omega \end{bmatrix} \begin{bmatrix} i_1 = 4mA \\ i_2 \end{bmatrix} = \begin{bmatrix} v_a \\ -12V \end{bmatrix} \qquad (8.12)$$

The second equation in Eq. (8.12) is

$$-16 \text{ V} + 5 \text{ k}\Omega \times i_2 = -12 \text{ V}$$

from which

$$i_2 = \frac{4}{5} \text{ mA} \tag{8.13}$$

We therefore see that

$$v_b = 1 \text{ k}\Omega \times i_2 = \frac{4}{5} \text{ V} \tag{8.14}$$

and

$$v_c = 2 \text{ k}\Omega \times \left(4 \text{ mA} - i_2\right) = 6.4 \text{ V} \tag{8.15}$$

Therefore,

$$v_{th} = v_{OC} = v_b - v_c = -5.6 \text{ V} \tag{8.16}$$

We can now find the Thevenin resistance by setting all sources to zero as shown in Fig. 8.13. Note that the equivalent resistance is just a 2 kΩ resistor in parallel with a 3 kΩ (2 kΩ+1 kΩ) resistor. Therefore,

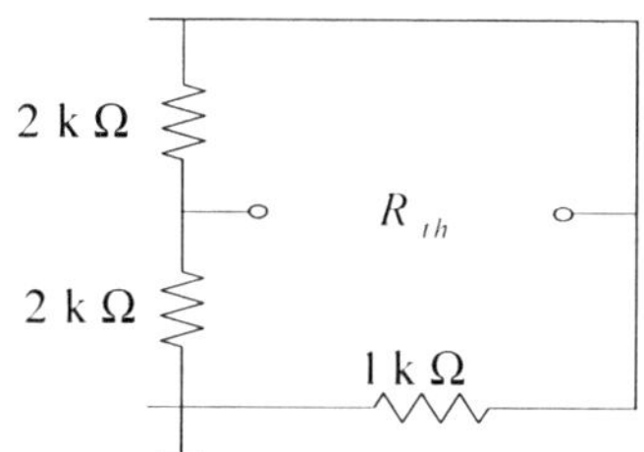

Figure 8.13 Example 48 – Finding R_{th}

$$R_{th} = \frac{2 \text{ k}\Omega \times 3 \text{ k}\Omega}{2 \text{ k}\Omega + 3 \text{ k}\Omega} = \frac{6}{5} \text{ k}\Omega = 1.2 \text{ k}\Omega \tag{8.17}$$

The Thevenin equivalent circuit is shown in Fig. 8.14.

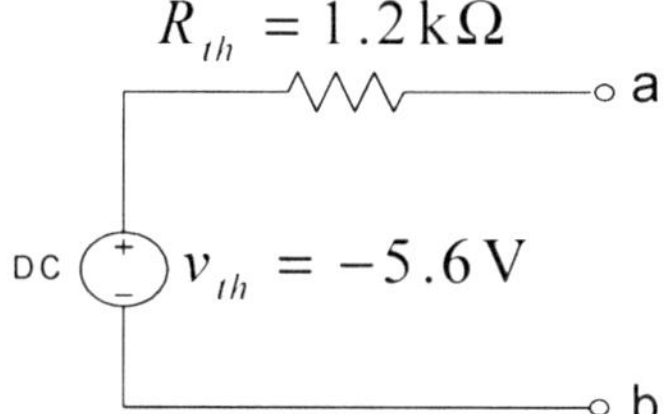

Figure 8.14 Example 48 - Thevenin equivalent circuit

To find v_0 in Fig. 8.11, we replace the 1 kΩ load resistor in Fig. 8.14 to obtain the circuit shown in Fig. 8.15. This is just a voltage divider from which we can see that

$$v_0 = \frac{1 \text{ k}\Omega}{1 \text{ k}\Omega + 1.2 \text{ k}\Omega}\left(-5.6 \text{ V}\right) = -2.5455 \text{ V} \tag{8.18}$$

which agrees with our result in Eq. (8.4) using superposition. The calculations for this example are given in Matlab Example 48.

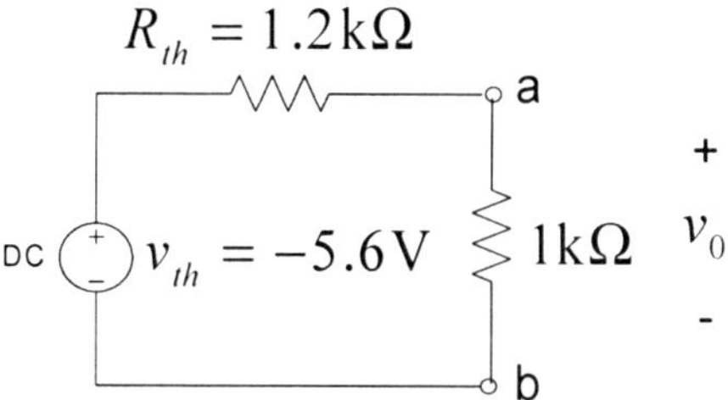

Figure 8.15 Example 48 – Finding v_0

Matlab Example 48

```
>> i2 = 4/5
i2 =
     0.8000

>> vb = 1*i2
vb =
     0.8000

>> vc = 2*(4-i2)
vc =
     6.4000

>> vth = vb - vc
vth =
    -5.6000

>> Rth = (2*3)/(2+3)
Rth =
     1.2000

>> v0 = (1/(1+1.2))*vth
v0 =
    -2.5455
```

Example 49 – Norton's Theorem

We saw in Fig. 8.10 that the Norton equivalent circuit is just a source transformation of the Thevenin equivalent circuit and vice versa. Therefore, once we have found the Thevenin equivalent circuit as we did in Example 48, we can draw the corresponding Norton equivalent circuit by using our rules for a source transformation. However, we can also find the Norton current directly by finding the short circuit current in our circuit. Let's do that for the same circuit as in Example 48 by drawing a short circuit across the load terminals in Fig. 8.12 to obtain the circuit shown in Fig. 8.16. We will then be able to verify our Thevenin results by computing v_0 using the Norton equivalent circuit.

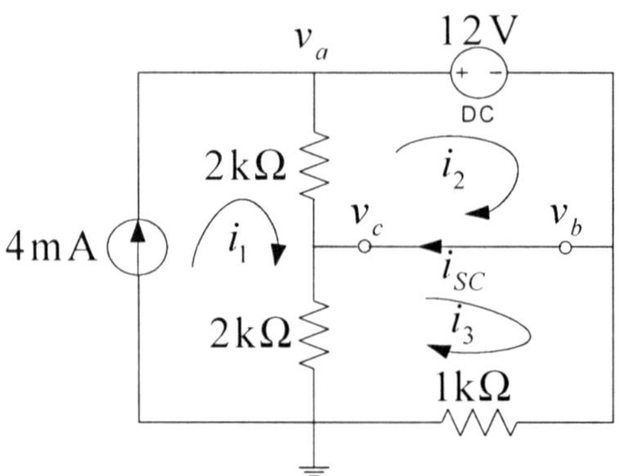

Figure 8.16 Example 49 – Finding i_{SC}

We need to find the mesh currents i_2 and i_3 in Fig. 8.16 from which we can calculate the short circuit current i_{SC}. Let's write the mesh equations by inspection.

$$\begin{bmatrix} 4\ \text{k}\Omega & -2\ \text{k}\Omega & -2\ \text{k}\Omega \\ -2\ \text{k}\Omega & 2\ \text{k}\Omega & 0 \\ -2\ \text{k}\Omega & 0 & 3\ \text{k}\Omega \end{bmatrix} \begin{bmatrix} i_1 = 4\ \text{mA} \\ i_2 \\ i_3 \end{bmatrix} = \begin{bmatrix} v_a \\ -12\ \text{V} \\ 0 \end{bmatrix} \qquad (8.19)$$

We could rewrite the above equations so that all the unknowns are on the left-hand side of the equation and use Matlab to find v_a, i_1, and i_3. Another way to solve the problem is to recognize that the second and third equations in Eq. (8.19) can be written as

$$-8\ \text{V} + 2\ \text{k}\Omega \times i_2 = -12\ \text{V} \qquad (8.20)$$

and

$$-8\ \text{V} + 3\ \text{k}\Omega \times i_3 = 0 \qquad (8.21)$$

From Eq. (8.20)

$$i_2 = -2\ \text{mA} \qquad (8.22)$$

and from Eq. (8.21)

$$i_3 = \frac{8}{3}\ \text{mA} = 2.6667\ \text{mA} \qquad (8.23)$$

Therefore,

$$i_{SC} = i_2 - i_3 = -4.6667\ \text{mA} \qquad (8.24)$$

Recall that the Norton resistance is the same as the Thevenin resistance, which we found in Eq. 8.17 to be 1.2 kΩ. The Norton equivalent circuit is shown in Fig. 8.17.

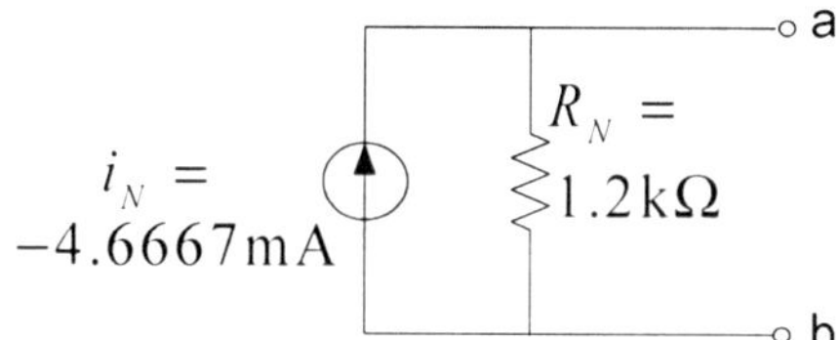

Figure 8.17 Example 49 - Norton equivalent circuit

To find v_0 in Fig. 8.11, we replace the 1 kΩ load resistor in Fig. 8.17 to obtain the circuit shown in Fig. 8.14. The voltage v_0 across the two parallel resistors is then given by

$$v_0 = \frac{1\ \text{k}\Omega \times 1.2\ \text{k}\Omega}{1\ \text{k}\Omega + 1.2\ \text{k}\Omega}(-4.6667\ \text{mA}) = -2.5455\ \text{V} \qquad (8.25)$$

which agrees with our result in Eq. (8.18) using Thevenin's theorem. The calculations for this example are given in Matlab Example 49.

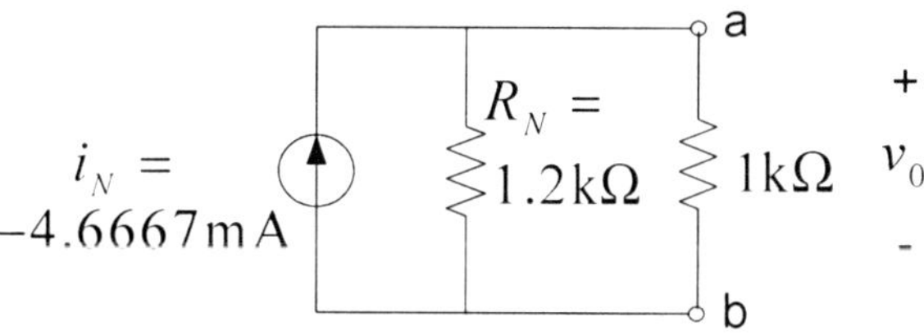

Figure 8.18 Example 49 – Finding v_0

Matlab Example 49

```
>> i2 = (-12+8)/2
i2 =
    -2

>> i3 = 8/3
i3 =
    2.6667

>> isc = i2 - i3
isc =
   -4.6667

>> v0 = ((1*1.2)/(1+1.2))*isc
v0 =
   -2.5455
```

Example 50 – AC Thevenin's Theorem

Use Thevenin's theorem to find the phasor current I_0 in the circuit shown in Fig. 7.10. This is the same circuit we solved in Example 45 using nodal analysis.

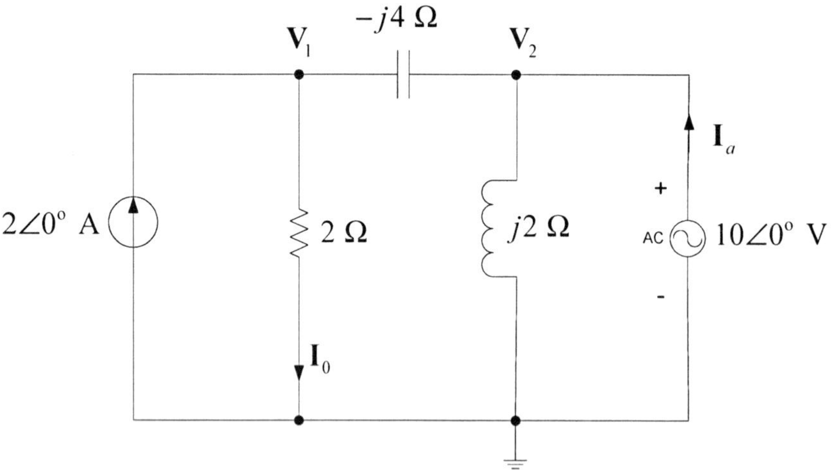

Figure 8.19 Circuit for Example 50

We begin by removing the 2 Ω resistor and finding the open circuit voltage V_{oc} as shown in Fig. 8.20.   Note that the current through the capacitor is 2 amps from the current source. Thus,

$$V_{oc} = V_C + V_2 = 2(-j4) + 10 = 10 - j8 \qquad (8.26)$$

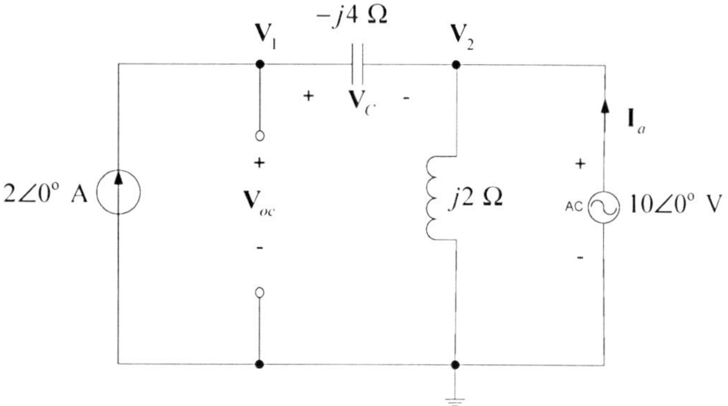

Figure 8.20 Finding the open circuit voltage $\mathbf{V}_{oc}$

Setting the values of all sources to zero in Fig. 8.21 we see that the Thevenin impedance is

$$\mathbf{Z}_{oc} = -j4 \ \Omega \qquad (8.27)$$

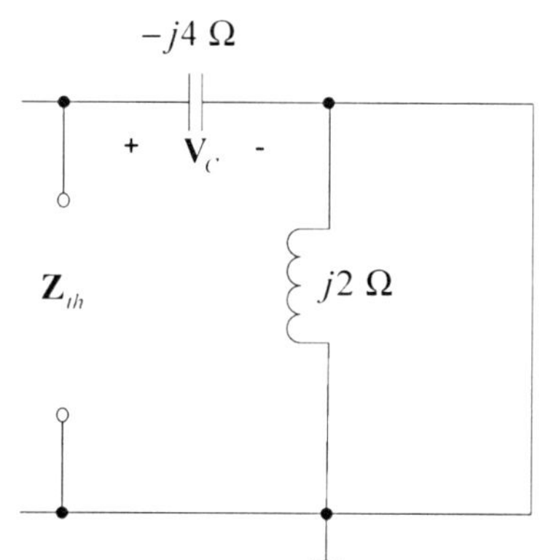

Figure 8.21 Finding the Thevenin impedance $\mathbf{Z}_{th}$

The resulting Thevenin equivalent circuit is shown in Fig. 8.22. By replacing the 2 Ω resistor in Fig. 8.23 we can immediately find the current $\mathbf{I}_0$ as

$$\mathbf{I}_0 = \frac{\mathbf{V}_{th}}{\mathbf{Z}_{th} + 2\Omega} = \frac{10 - j8}{2 - j4} = 2.867\angle24.78^\circ \text{ A} \qquad (8.28)$$

which agrees with the result we found in Example 45. The calculations are shown in Matlab Example 50.

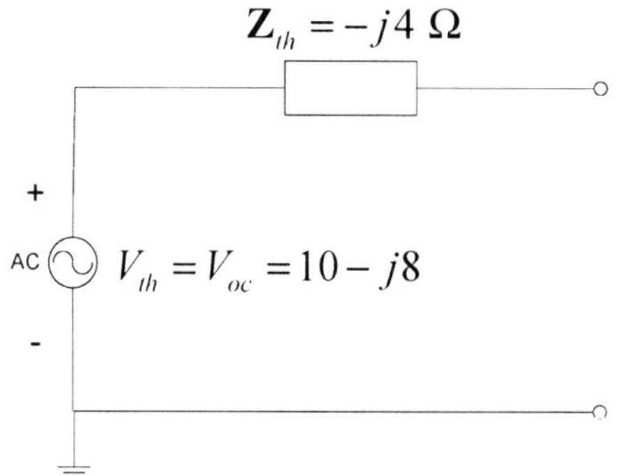

Figure 8.22 Thevenin equivalent circuit

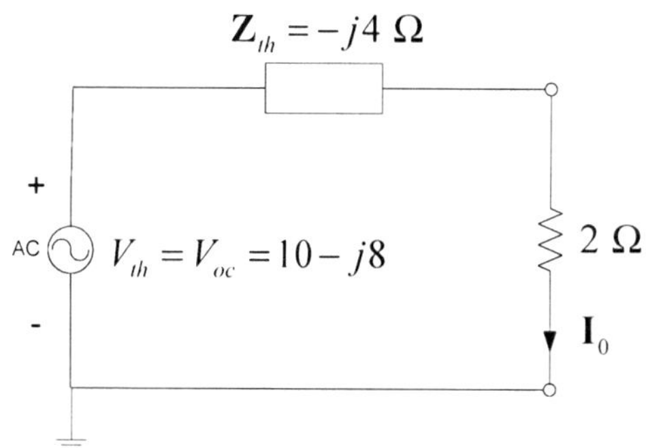

Figure 8.23 Solving for the current $\mathbf{I}_0$

Matlab Example 50

```
>> Vth = 2*(-j*4) + 10
Vth =
   10.0000 - 8.0000i

>> I0 = (10 - 8j)/(2 - 4j)
I0 =
   2.6000 + 1.2000i

>> magI0 = abs(I0)
magI0 =
     2.8636

>> angI0 = angle(I0)*180/pi
angI0 =
    24.7751
```

8.3 Maximum Power Transfer

In all practical cases, energy sources have non-zero internal resistance.   Thus, there are losses inherent in any real source. Also, in most cases, the aim of an energy source is to provide power to a load. Given a circuit with a known internal resistance, what is the resistance of the load that will result in the maximum power being delivered to the load?

Consider the source to be modeled by its Thevenin equivalent circuit as shown in Fig. 8.24.

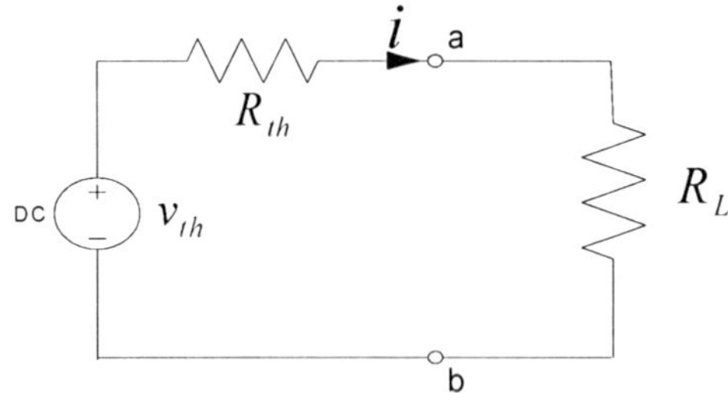

Figure 8.24 Thevenin equivalent circuit driving a load R_L

The power delivered to the load (absorbed by R_L) is

$$p = i^2 R_L = \left[v_{th} / \left(R_{th} + R_L \right) \right]^2 R_L \tag{8.29}$$

This power is maximum when $\partial p / \partial R_L = 0$. Differentiating Eq. (8.29), we obtain

$$\frac{\partial p}{\partial R_L} = v_{th}^2 \left[\left(R_{th} + R_L \right)^{-2} - 2R_L \left(R_{th} + R_L \right)^{-3} \right] = 0 \tag{8.30}$$

The term in the square bracket in Eq. (8.30) will be zero if

$$R_{th} + R_L = 2R_L \tag{8.31}$$

or

$$R_L = R_{th} \tag{8.32}$$

Thus, maximum power transfer takes place when the resistance of the load equals the Thevenin resistance R_{th}. If we set $R_L = R_{th}$ in Eq. (8.29), we see that the maximum power delivered to the load is

$$p_{max} = \left[v_{th} / (2R_{th}) \right]^2 R_{th} = v_{th}^2 / 4R_{th} \tag{8.33}$$

The total power supplied by the voltage source will be $v_{th}^2 / 2R_{th}$. Thus, we see from Eq. (8.33) that, at best, one-half of the power is dissipated in the internal resistance and one-half in the load.

Example 51 – Maximum Power Transfer

In Fig. 8.24, let $v_{th} = 10$ V and $R_{th} = 50\ \Omega$. Using Eq. (8.29), plot the power dissipated in the load resistor R_L for values of R_L between 0 and 200 Ω.

The solution is given in Matlab Example 51. The resulting plot is shown in Fig. 8.16. Note that the maximum power occurs for a value of $R_L = R_{th} = 50\ \Omega$.

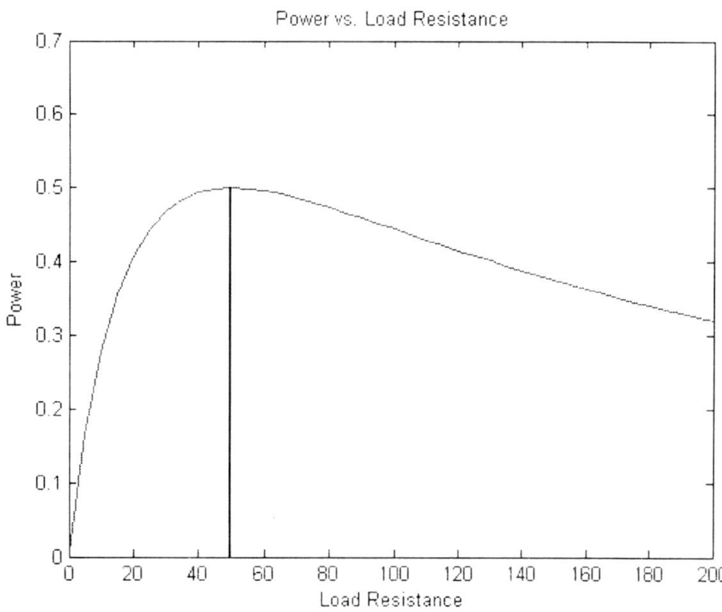

Figure 8.25 Plot generated by Matlab Example 51

Matlab Example 51

```
>> vth = 10;
>> rth = 50;
>> rl = 0:5:200
rl =
  Columns 1 through 12
     0     5    10    15    20    25    30    35    40    45    50    55
  Columns 13 through 24
    60    65    70    75    80    85    90    95   100   105   110   115
  Columns 25 through 36
   120   125   130   135   140   145   150   155   160   165   170   175
  Columns 37 through 41
   180   185   190   195   200

>> p = ((vth./(rth+rl)).^2).*rl
p =
  Columns 1 through 7
        0    0.1653    0.2778    0.3550    0.4082    0.4444    0.4688
  Columns 8 through 14
   0.4844    0.4938    0.4986    0.5000    0.4989    0.4959    0.4915
  Columns 15 through 21
   0.4861    0.4800    0.4734    0.4664    0.4592    0.4518    0.4444
  Columns 22 through 28
   0.4370    0.4297    0.4224    0.4152    0.4082    0.4012    0.3944
  Columns 29 through 35
   0.3878    0.3813    0.3750    0.3688    0.3628    0.3569    0.3512
  Columns 36 through 41
   0.3457    0.3403    0.3350    0.3299    0.3249    0.3200

>> plot(rl,p)
```

Problems

8.1   In the circuit shown at the right, find the current i_0 through the 3Ω resistor using superposition. Compare your result with that in Example 44.

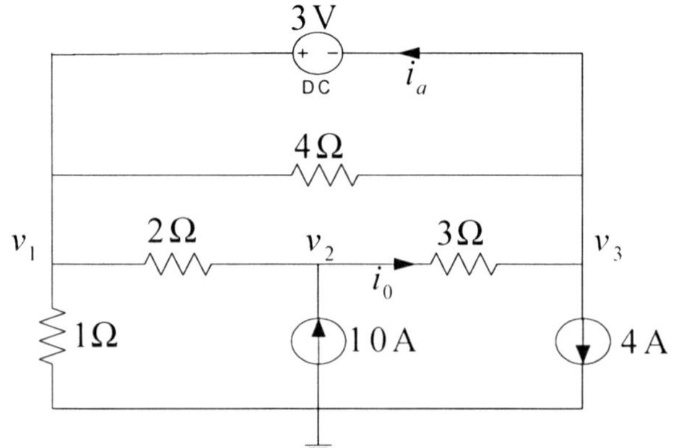

8.2   Repeat Problem 8.1 using Thevenin's theorem.

8.3   Repeat Problem 8.1 using Norton's theorem.

Chapter 9

Power and Transformers

The electrical power used in your home and the nation's factories is obviously of great importance. In this chapter, you will learn how to calculate the power that is required to be provided by the sources in AC circuits. You will also learn how an ideal transformer can transform voltages and currents in AC circuits.

9.1 Instantaneous and Average Power

Suppose that a voltage or current source produces a voltage and current at its terminals given by

$$v(t) = V_M \cos(\omega t + \theta_v) \qquad (9.1)$$

and

$$i(t) = I_M \cos(\omega t + \theta_i) \qquad (9.2)$$

Then the instantaneous power delivered to a circuit connected to these terminals will be

$$p(t) = v(t)i(t) = V_M I_M \cos(\omega t + \theta_v)\cos(\omega t + \theta_i) \qquad (9.3)$$

Using the trigonometric identity

$$\cos A \cos B = \frac{1}{2}\left[\cos(A - B) + \cos(A + B)\right] \qquad (9.4)$$

we can write Eq. (9.3) as

$$p(t) = \frac{V_M I_M}{2}\left[\cos(\theta_v - \theta_i) + \cos(2\omega t + \theta_v + \theta_i)\right] \qquad (9.5)$$

The first term in Eq. (9.5) is a constant that depends on $\cos(\theta_v - \theta_i)$ and the second term is a sinusoidal signal at twice the frequency of the voltage and current. Note that this agrees with the results we obtained in Fig. 2.18 of Example 4 and Fig. 2.27 of Example 6.

The average power is defined as the average of $p(t)$ over one period $T = 2\pi/\omega$. Thus,

$$P_{avg} = \frac{1}{T}\int_{t_0}^{t_0+T} p(t)\,dt$$

$$= \frac{1}{T}\int_{t_0}^{t_0+T} V_M I_M \cos(\omega t + \theta_v)\cos(\omega t + \theta_i)\,dt \tag{9.6}$$

Using Eq. (9.4), we can write Eq. (9.6) as

$$P_{avg} = \frac{1}{T}\int_{t_0}^{t_0+T} \frac{V_M I_M}{2}\left[\cos(\theta_v - \theta_i) + \cos(2\omega t + \theta_v + \theta_i)\right]dt \tag{9.7}$$

Note that the second term in Eq. (9.7) is a sinusiodal signal whose integral over one period will be zero. The first term is just a constant and so Eq. (9.7) integrates to

$$P_{avg} = \tfrac{1}{2} V_M I_M \cos(\theta_v - \theta_i) \tag{9.8}$$

For a purely resistive circuit, $\theta_v - \theta_i = 0$, and Eq. (9.8) reduces to

$$P_{avg} = \tfrac{1}{2} V_M I_M \qquad \text{resistive circuit} \tag{9.9}$$

For a purely reactive circuit, $\theta_v - \theta_i = \pm 90°$, and Eq. (9.8) reduces to

$$P_{avg} = \tfrac{1}{2} V_M I_M \cos(90°) = 0 \qquad \text{reactive circuit} \tag{9.10}$$

Recall from Eq. (4.12) in Section 4.2 that the amplitudes V_M and I_M are related to the corresponding RMS values by

$$V_M = V_{rms}\sqrt{2} \qquad\qquad I_M = I_{rms}\sqrt{2} \tag{9.11}$$

If we substitute V_M and I_M from Eq. (9.11) into Eq. (9.10), we see that we can write the average power in terms of the RMS voltages and currents as

$$P_{avg} = V_{rms} I_{rms} \cos(\theta_v - \theta_i) \tag{9.12}$$

In Eq. (9.12) the factor $\cos(\theta_v - \theta_i)$ is called the *power factor*.

Example 52 – Instantaneous and Average Power

Consider the R-L circuit from Fig. 4.15 in Example 16 that is reshown in Fig. 9.1. If R = 600 Ω, L = 1 mH, V_M = 2 V, and the frequency f = 100 kHz, plot the voltage $v(t) = V_M \cos \omega t$ and the instantaneous power $p(t)$ as a function of time for two cycles of the voltage. Calculate the average power.

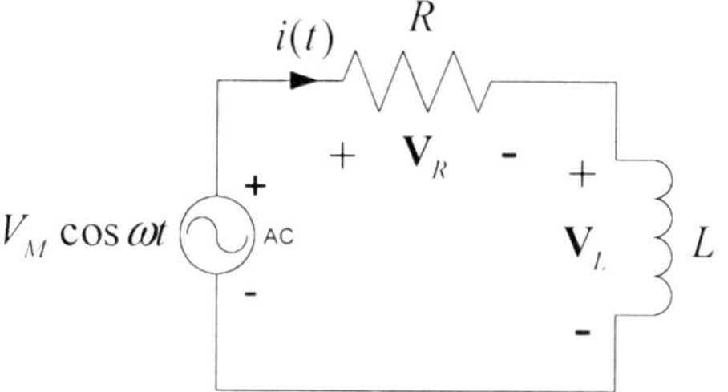

Figure 9.1 Example 52

The instantaneous power $p(t)$ is given by Eq. (9.5) and the average power P_{avg} is given by Eq. (9.8). Listing 9.1 is a program for a Matlab function $pRL(R,L,V,f)$ that will plot the voltage and the instantaneous power as a function of time for two cycles of the voltage. Input parameters are the values of resistance, inductance, voltage magnitude, and frequency. Matlab Example 52 shows a run to this program that produces the plots shown in Fig. 9.2. Note, in Listing 9.1, that we have multiplied the equation for $p(t)$ by 10^3 so that the scale of power values in Fig. 9.2 will be milliwatts. We did the same for the calculation of average power.

Listing 9.1 pRL.m

```
function pRL(R,L,V,f)
% The voltage V is across R and L in series
% plot voltage, current and instantaneous power
%    as a function of time for 2 cycles of V
% calculate average power
% R = resistance in ohms
% L = inductance in henrys
% V = voltage in volts
% f = frequency in Hz
% pRL(R,L,V,f)
%
t = linspace(0,2/f,100);
w = 2*pi.*f;
v = V*cos(w.*t);
x = w*L
thetai = -atan(x/R);
thetaid = thetai*180/pi
Im = V/sqrt(R^2+x^2)
i = Im*cos(w.*t+thetai)*10^3;
p = (V*Im/2)*(cos(-thetai)+cos(2*w.*t+thetai))*10^3;
Pavg = 0.5*V*Im*cos(-thetai)*10^3
curves = [v;i;p];
plot(t,curves)
```

Matlab Example 52

```
>> help pRL
  The voltage V is across R and L in series
  plot voltage, current and instantaneous power
  as a function of time for 2 cycles of V
  calculate average power
  R = resistance in ohms
  L = inductance in henrys
  V = voltage in volts
  f = frequency in Hz
  pRL(R,L,V,f)

>> R = 600;
>> L = 0.001;
>> V = 2;
>> f = 10^5;
>> pRL(R,L,V,f)

x =
  628.3185

thetaid =
  -46.3207

Im =
    0.0023

Pavg =
    1.5899
```

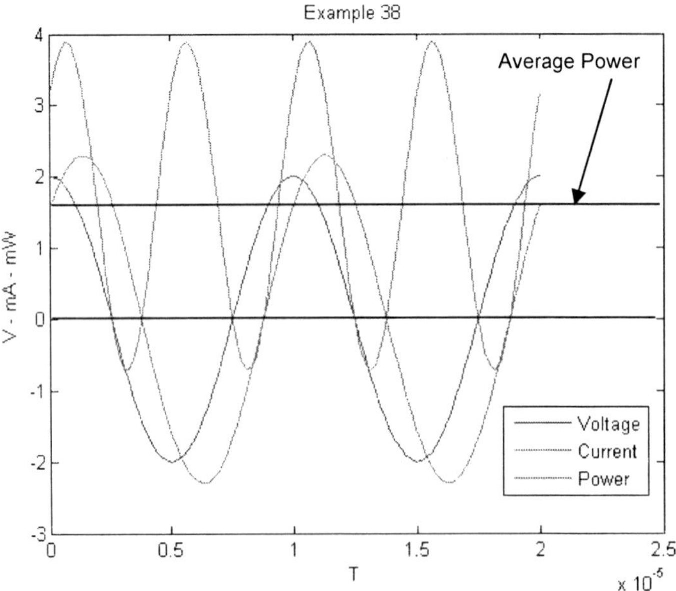

Figure 9.2 Plot resulting from Matlab Example 52

Note, in Fig. 9.2, that the instantaneous power varies at twice the frequency of the voltage waveform. The instantaneous power is positive when the voltage and current have the same sign and is negative when the voltage and current have opposite signs. When the instantaneous power is positive, power is being absorbed by the circuit. When the instantaneous power is negative, power is being transferred from the circuit to the source. In this case, that power comes from energy stored in the magnetic field of the inductor. Compare Fig. 9.2 with Fig. 2.27 of Example 6 in Chapter 2. Note that when the resistance is zero, as in Fig. 2.27, the current and voltage are $90°$ out of phase and the average power is zero. On the other hand, if the inductance is zero, then the phase between the voltage and current in Fig. 9.2 will go to zero and the average power will be given by $P_{avg} = (1/2)V_M I_M = V_{rms} I_{rms}$ (see Eqs. (9.10) and (9.12)), since $\cos 0° = 1$.

Example 53 – Voltages and Average Power in Resonant Circuits

Consider the resonant circuit from Example 21 in Chapter 5 that is redrawn in Fig. 9.3. Let R = 600 Ω, C = 2.653 nF, L = 9.5 mH, V_M = 1 V, and f = 10^5 Hz. Plot the voltages $v_R(t)$, $v_L(t)$, $v_C(t)$, and the instantaneous power $p(t)$ as a function of time for two cycles of the input voltage.

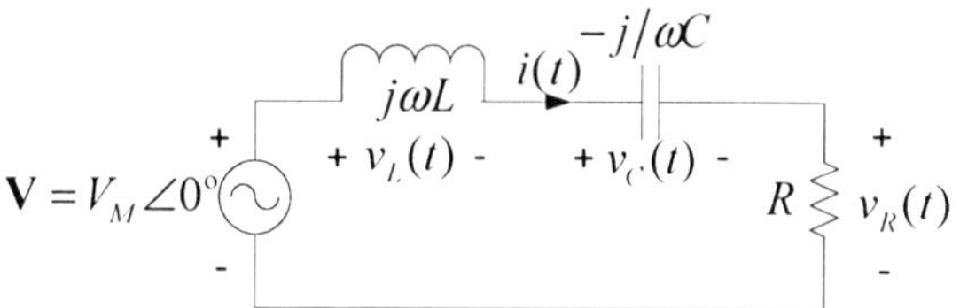

Figure 9.3 Example 53

The total impedance of the circuit is

$$\mathbf{Z} = R + j\omega L - j/\omega C = R + jX_L + jX_C = R + jX \qquad (9.13)$$

where $X_L = \omega L$, $X_C = -1/\omega C$ and $X = X_L + X_C$. The phasor current through the circuit is

$$\mathbf{I} = \frac{\mathbf{V}}{\mathbf{Z}} = \frac{V_M}{R + jX} = I_M \angle \theta_i \qquad (9.14)$$

where

$$I_M = \frac{V_M}{\sqrt{R^2 + X^2}} \qquad (9.15)$$

and

$$\theta_i = -\tan^{-1}(X/R) \qquad (9.16)$$

These values are calculated in the Matlab program in Listing 9.2. The voltage $v_R(t)$ across the resistor is given by

$$v_R(t) = \mathrm{Re}\left\{ I_M e^{j\theta_i} e^{j\omega t} R \right\} = I_M R \cos(\omega t + \theta_i) \tag{9.17}$$

Similarly, the voltages across the inductor and capacitor are given by

$$v_L(t) = \mathrm{Re}\left\{ I_M e^{j\theta_i} e^{j\omega t} j\omega L \right\} = I_M \omega L \cos(\omega t + \theta_i + \pi/2) \tag{9.18}$$

and

$$v_C(t) = \mathrm{Re}\left\{ I_M e^{j\theta_i} e^{j\omega t} (-j/\omega C) \right\} = I_M (1/\omega C) \cos(\omega t + \theta_i - \pi/2) \tag{9.19}$$

Matlab Example 53 shows a run of this program that produces the plots shown in Fig. 9.4.

Listing 9.2 pRLC.m

```
function pRLC(R,L,C,V,f)
% The voltage V is across RLC in series
% plot voltages across R,L,C and instantaneous power
%    as a function of time for 2 cycles of V
% calculate average power
% R = resistance in ohms
% L = inductance in henrys
% C = capacitance in farads
% V = voltage in volts
% f = frequency in Hz
% pRL(R,L,C,V,f)
%
t = linspace(0,2/f,100);
w = 2*pi.*f;
v = V*cos(w.*t);
xl = w*L
xc = -1/(w*C)
x = w*L+1/(w*C)
thetai = -atan(x/R);
thetaid = thetai*180/pi
Im = V/sqrt(R^2+x^2)
i = Im*cos(w.*t+thetai);
vr = Im*R*cos(w.*t+thetai);
vl = Im*w*L*cos(w.*t+thetai+pi/2);
vc = Im*(1/(w*C))*cos(w.*t+thetai-pi/2);
p = (V*Im/2)*(cos(-thetai)+cos(2*w.*t+thetai))*10^3;
Pavg = 0.5*V*Im*cos(-thetai)*10^3
curves = [vr;vl;vc;p];
plot(t,curves)
```

Matlab Example 53

```
>> help pRLC
  The voltage V is across RLC in series
  plot voltage, current and instantaneous power
    as a function of time for 2 cycles of V
  calculate average power
  R = resistance in ohms
  L = inductance in henrys
  C = capacitance in farads
  V = voltage in volts
  f = frequency in Hz
  pRL(R,L,C,V,f)
>> R = 600;
>> L = 0.0095;
>> C = 0.2653*10^-9;
>> f = 10^5;
>> V = 1;
>> pRLC(R,L,C,V,f)

xl =
  5.9690e+003
xc =
  -5.9991e+003
x =
  -30.0295
thetaid =
    2.8652
Im =
    0.0017
Pavg =
    0.8313
```

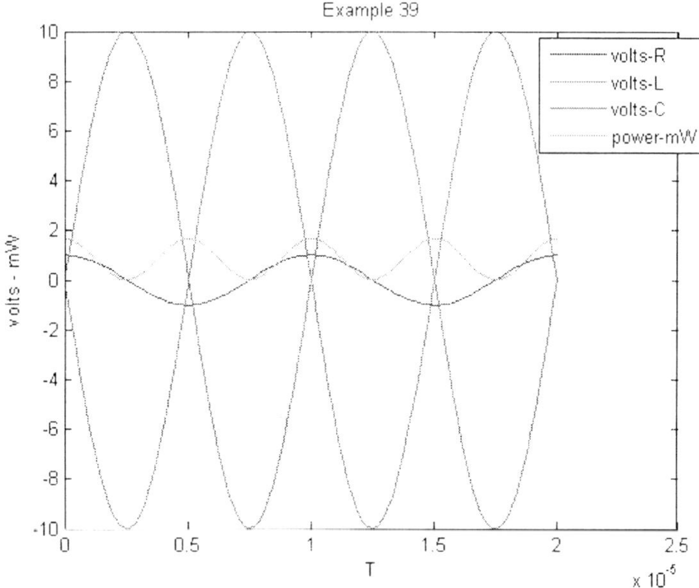

Figure 9.4 Plot resulting from Matlab Example 53

This circuit is operating at the resonant frequency 10^5 Hz shown in Fig. 8.16 and a remarkable result is shown in Fig. 9.4. The input voltage is a 1 V cosine wave and the voltage across the resistor is the same 1 V cosine wave. Thus, the total voltage across both the inductor and capacitor is always zero. However, the actual voltage across the capacitor is a 10 V sine wave and the actual voltage across the inductor is a 10 V negative sine wave! Notice that the instantaneous power (and instantaneous current) is just the instantaneous power you would expect from the 1 V sinusoidal voltage source connected across a 600 Ω resistor. However, the capacitor and inductor must be able to withstand the potentially high voltages that may occur in series resonant circuits. Potentially high currents can occur in parallel resonant circuits (see Problem 9.1).

9.2 Ideal Transformer

We saw in Chapter 1 that current flowing in a wire produces a circular magnetic field around the wire. The direction of the magnetic field is given by the right-hand rule where, if the thumb on your right hand points in the direction of the current, your fingers will show the direction of the magnetic field. When you wrap a wire around a core as shown in Fig. 9.5, the magnetic fields created by each loop of the coil will all point in the same direction and add up to produce a magnetic field that points upward within the coil in Fig. 9.5. If each loop in the coil produces a magnetic flux ϕ (with units of webers), then a coil with N loops will have a total magnetic flux equal to $N\phi$. This total magnetic flux will be proportional to the current I in the wire. The constant of proportionality is the inductance L. We can therefore write

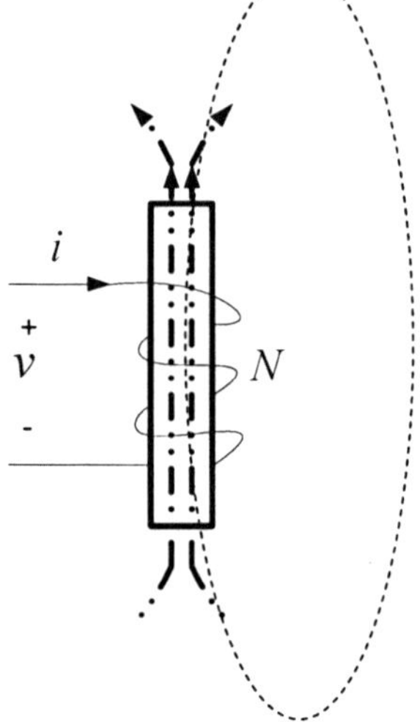

$$N\phi = Li \qquad (9.20)$$

Faraday's law states that the changing magnetic field within the coil will induce a voltage $v(t)$ across the coil given by

$$v(t) = N\frac{d\phi}{dt} = L\frac{di}{dt} \qquad (9.21)$$

Figure 9.5
Current in a coil produces a magnetic field

This agrees with our equation for the voltage across an inductor from Section 2.5.

Now suppose we wrap a coil containing N_1 turns around the left, or primary, side of a rectangular iron core as shown in Fig. 9.6. The iron core will cause the magnetic flux generated by the current i_1 to be confined to the core and circulate around the entire core. From Eq. 9.21 the voltage $v_1(t)$ in Fig. 9.6 is given by

$$v_1(t) = N_1 \frac{d\phi}{dt} \qquad (9.22)$$

If we now wrap a second coil containing N_2 turns on the right, or secondary, side of the iron core as shown in Fig. 9.6, the changing magnetic flux $d\phi/dt$ will induce a voltage $v_2(t)$ given by

$$v_2(t) = N_2 \frac{d\phi}{dt} \qquad (9.23)$$

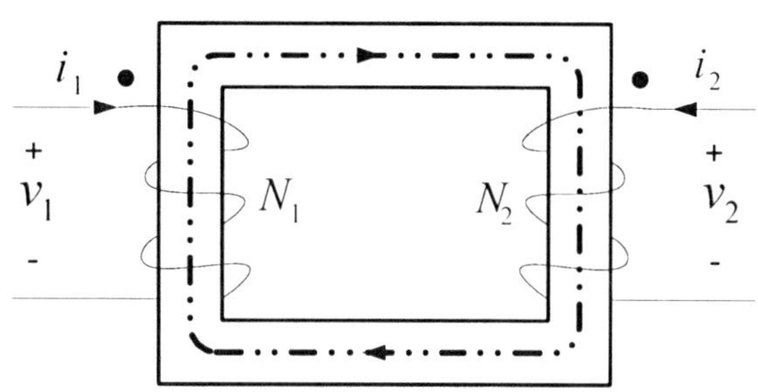

Figure 9.6
Diagram of an ideal transformer

Note that, if the current i_2 shown in Fig. 9.6 enters the coil N_2, the generated flux will add to the flux from coil N_1. The two dots shown on the coils indicate which way the coils are wound. If current enters both dots the generated fluxes add in the same direction.

If we divide Eq. (9.22) by Eq. (9.23), we obtain

$$\frac{v_1}{v_2} = \frac{N_1}{N_2} \frac{\dfrac{d\phi}{dt}}{\dfrac{d\phi}{dt}} = \frac{N_1}{N_2} \qquad (9.24)$$

from which

$$v_2 = \frac{N_2}{N_1} v_1 \qquad (9.25)$$

The turns ratio, n, is defined by

$$n = \frac{N_2}{N_1} \qquad (9.26)$$

Note that if $N_2 > N_1$, then the output voltage v_2 will be greater than the input voltage v_1 and we have a step-up transformer. If $N_2 < N_1$, then the output voltage v_2 will be less than the input voltage v_1 and we have a step-down transformer.

Consider a voltage source v1 connected to the primary side of a transformer and a resistive load connected to the secondary side as shown in Fig. 9.7. An ideal transformer is lossless so that the input power $P_1 = v_1 i_1$ will be equal to the output power $P_2 = v_2 i_2$. We can therefore write

$$v_2 i_2 = v_1 i_1 \qquad (9.27)$$

Substituting Eq. (9.25) into Eq. (9.27), we can write

$$\frac{i_1}{i_2} = \frac{v_2}{v_1} = \frac{N_2}{N_1}$$

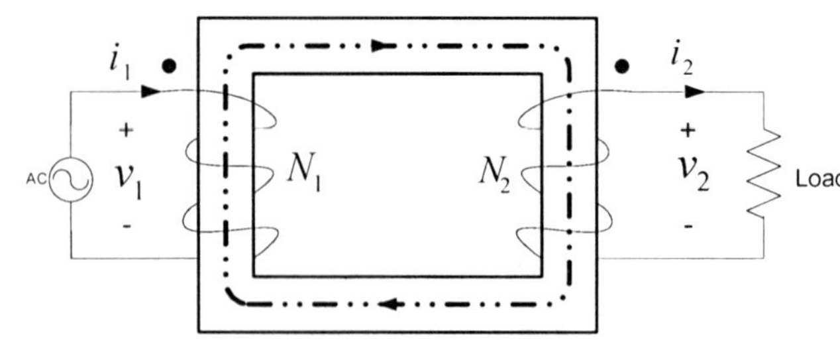

Figure 9.7
An ideal transformer is lossless

from which

$$i_2 = \frac{N_1}{N_2} i_1 \qquad (9.28)$$

Consider the ideal transformer shown in Fig. 9.8 where the input and output phasor voltages and currents are indicated. The output load impedance will be given by

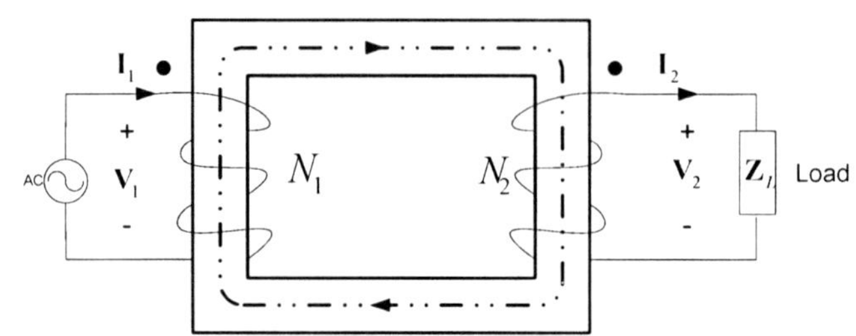

Figure 9.8
An ideal transformer

$$\mathbf{Z}_L = \frac{\mathbf{V}_2}{\mathbf{I}_2} \qquad (9.29)$$

From Eq. (9.28), we can write the input phasor current as

$$\mathbf{I}_1 = \frac{N_2}{N_1} \mathbf{I}_2 \qquad (9.30)$$

and, from Eq. (9.25), we can write the input phasor voltage as

$$\mathbf{V}_1 = \frac{N_1}{N_2} \mathbf{V}_2 \qquad (9.31)$$

From Eqs. (9.29) – (9.31), we can write the input impedance as

$$\mathbf{Z}_i = \frac{\mathbf{V}_1}{\mathbf{I}_1} = \left(\frac{N_1}{N_2}\right)^2 \mathbf{Z}_L \qquad (9.32)$$

or, in terms of the turns ratio n given by Eq. (9.26),

$$\mathbf{Z}_i = \frac{\mathbf{Z}_L}{n^2} \qquad (9.33)$$

Eqs. (9.32) and (9.33) show how the load impedance across the secondary of a transformer is reflected to the primary side of the transformer.

Example 54 – Transformer Network

Find the current $\mathbf{I}_1$ and the voltage $\mathbf{V}_0$ in the circuit shown in Fig. 9.9.

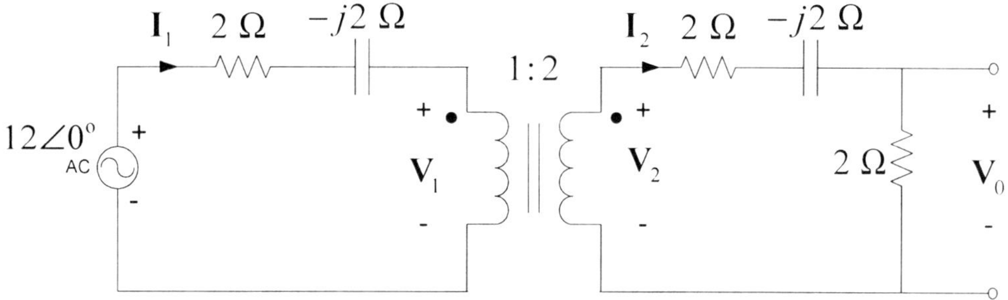

Figure 9.9 Circuit for Example 54

The total load impedance is

$$\mathbf{Z}_L = 4 - j2 \ \Omega$$

The turns ratio is $n = 2$; therefore, the reflected input impedance $\mathbf{Z}_i$ shown in Fig. 9.10 is

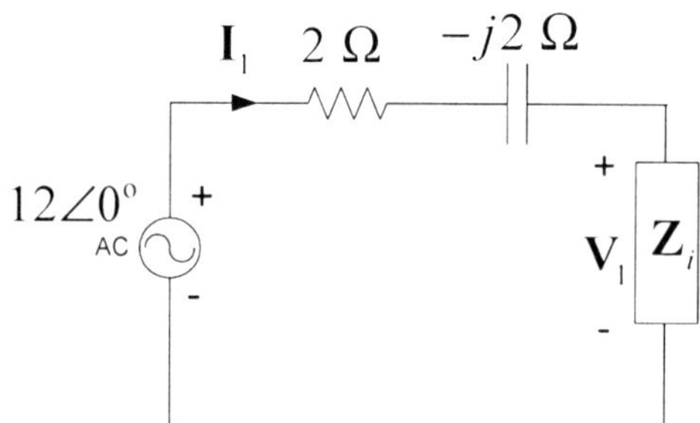

Figure 9.10 Example 54 – Step 2

$$\mathbf{Z}_i = \frac{\mathbf{Z}_L}{n^2} = 1 - j0.5$$

The input current $\mathbf{I}_1$ can then be found to be

$$\mathbf{I}_1 = \frac{12\angle 0^{\circ}}{2 - j2 + \mathbf{Z}_i} = 3.07\angle 39.81^{\circ} \ \text{A} \tag{9.34}$$

The input primary voltage $\mathbf{V}_1$ in Fig. 9.10 is then

$$\mathbf{V}_1 = \mathbf{I}_1\mathbf{Z}_i = 3.344 + j0.787$$

and the secondary voltage $\mathbf{V}_2$ in Fig. 9.9 is

$$\mathbf{V}_2 = n\mathbf{V}_1 = 6.689 + j1.574$$

The output voltage $\mathbf{V}_0$ can then be found by voltage division to be

$$\mathbf{V}_0 = \mathbf{V}_2 \frac{2}{4 - j2} = 3.07\angle 39.81^{\circ} \ \text{V} \tag{9.35}$$

Matlab Example 54 shows the calculations for Example 54.

Matlab Example 54

```
>> ZL = 4-2j
ZL =
   4.0000 - 2.0000i

>> n = 2
n =
     2

>> Zi = ZL/n^2
Zi =
   1.0000 - 0.5000i

>> I1 = 12/(2-2j+Zi)
I1 =
   2.3607 + 1.9672i

>> ampI1 = abs(I1)
ampI1 =
    3.0729

>> angI1 = angle(I1)*180/pi
angI1 =
   39.8056

>> V1 = I1*Zi
V1 =
   3.3443 + 0.7869i

>> V2 = n*V1
V2 =
   6.6885 + 1.5738i

>> V0 = V2*2/(4-2j)
V0 =
   2.3607 + 1.9672i

>> ampV0 = abs(V0)
ampV0 =
    3.0729

>> angV0 = angle(V0)*180/pi
angV0 =
   39.8056
```

9.3 Complex Power

A convenient way to understand average power in an AC circuit is by using the concept of *complex power* that is defined as

$$\mathbf{S} = \mathbf{V}_{rms}\mathbf{I}_{rms}^{*} \tag{9.36}$$

where I_{rms}^{*} = complex conjugate of I_{rms}. Thus, if

$$\mathbf{I}_{rms} = I_r + jI_i = I_{rms}\angle\theta_i \tag{9.37}$$

then

$$\mathbf{I}_{rms}^{*} = I_r - jI_i = I_{rms}\angle-\theta_i \tag{9.38}$$

From Eqs. (9.36) and (9.37), we can write

$$\mathbf{S} = V_{rms}\angle\theta_v I_{rms}\angle-\theta_i = V_{rms}I_{rms}\angle\left(\theta_v - \theta_i\right) \tag{9.39}$$

or

$$\mathbf{S} = P + jQ \tag{9.40}$$

where

$$P = \mathrm{Re}(\mathbf{S}) = V_{rms}I_{rms}\cos\left(\theta_v - \theta_i\right) \tag{9.41}$$

is the real, average power given by Eq. (9.12) and

$$Q = \mathrm{Im}(\mathbf{S}) = V_{rms}I_{rms}\sin\left(\theta_v - \theta_i\right) \tag{9.42}$$

is the imaginary, or *reactive*, or *quadrature*, power. The units of Q are given as *volt-ampere reactive* (VAR) to distinguish it from watts, the units of the real, average power P. The magnitude of the complex power $\mathbf{S}$ is given by

$$S = V_{rms}I_{rms} \tag{9.43}$$

and is called the *apparent power* with units *volt-amperes* (VA). Recall that the power factor PF is given by $\cos\left(\theta_v - \theta_i\right)$. Therefore, from Eqs. (9.41) and (9.43), we can write

$$\text{Power Factor}\ \ PF = \cos\left(\theta_v - \theta_i\right) = \frac{P}{S}$$

These results are summarized in the power triangle shown in Fig. 9.11. Recall that impedance is defined as

$$\mathbf{Z} = \frac{\mathbf{V}}{\mathbf{I}} = \frac{\mathbf{V}_{rms}}{\mathbf{I}_{rms}} = \frac{V_{rms}\angle\theta_v}{I_{rms}\angle\theta_i} = Z\angle\left(\theta_v - \theta_i\right) = R + jX \tag{9.44}$$

which can be represented by the impedance triangle in Fig. 9.12.

Figure 9.11 Power triangle   Figure 9.12 Impedance triangle

From Eq. (9.44), we can write

$$\mathbf{V}_{rms} = \mathbf{Z}\mathbf{I}_{rms} \tag{9.45}$$

Substituting Eq. (9.44) into Eq. (9.36), we obtain

$$\mathbf{S} = \mathbf{Z}\mathbf{I}_{rms}\mathbf{I}_{rms}^{*} = I_{rms}^{2}\mathbf{Z} = I_{rms}^{2}\left(R + jX\right) \tag{9.46}$$

Comparing Eqs. (9.46) and (9.40), we see that

$$P = I_{rms}^{2}R \tag{9.47}$$

and

$$Q = I_{rms}^{2}X \tag{9.48}$$

Power Factor Correction

Most industrial plants have machinery that can be modeled as an inductive load as shown in Fig. 9.13. Because the load is inductive, the current through the load will lag the voltage across the load by an angle θ_1 as shown in Fig. 9.14. We say that the load has a *lagging power factor* equal to $\cos\theta_1$.

Now only the component of $\mathbf{I}_{load}$ in the direction of $\mathbf{V}_{load}$ contributes to the real power absorbed by the load (see Eq. (9.39)). However, the total magnitude of $\mathbf{I}_{load}$ must be supplied by the voltage generating plant and this total current will contribute to power lost in the line resistance R_{line} shown in Fig. 9.13. For this reason, power companies charge industrial customers for use of *reactive power*, that is, they charge for low power factors. It is therefore in an industrial user's interest to keep the power factor of their plant as close to 1 as possible.

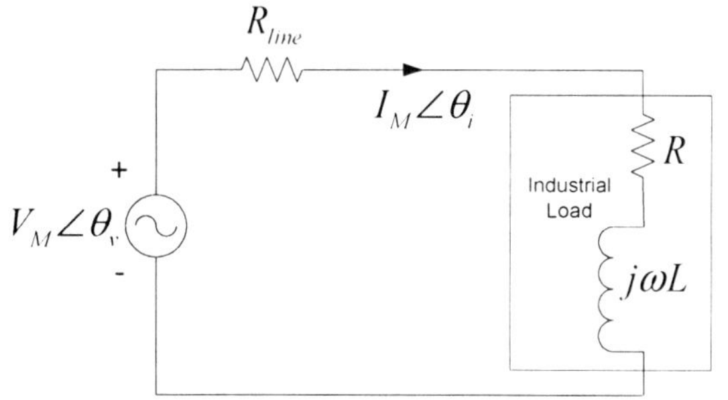

Figure 9.13 Modeling an industrial load

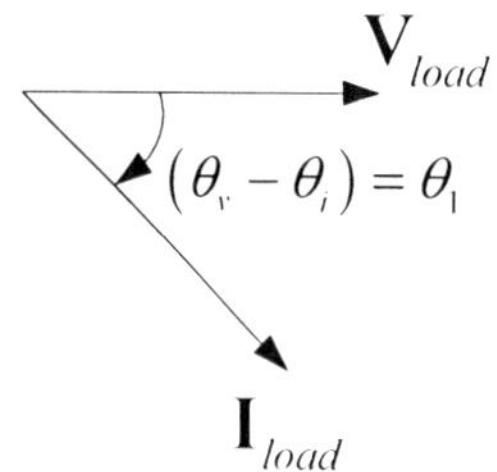

Figure 9.14 Power factor is $\cos\theta_1$

The easiest way to increase the power factor of an inductive load is to add a capacitor across the load as shown in Fig. 9.15. The current through the capacitor will lead the load voltage by $90°$ as shown in Fig. 9.16. If we want to keep the load current the same so that the same power will be delivered to the load, then the new line current will be the sum of the load current $\mathbf{I}_{load}$ and the capacitor current $\mathbf{I}_C$ as shown in Fig. 9.16. Note that the new power factor is $\cos\theta_2$ and the magnitude of the line current is smaller than in Fig. 9.13. Therefore, the power lost in the transmission line is smaller.

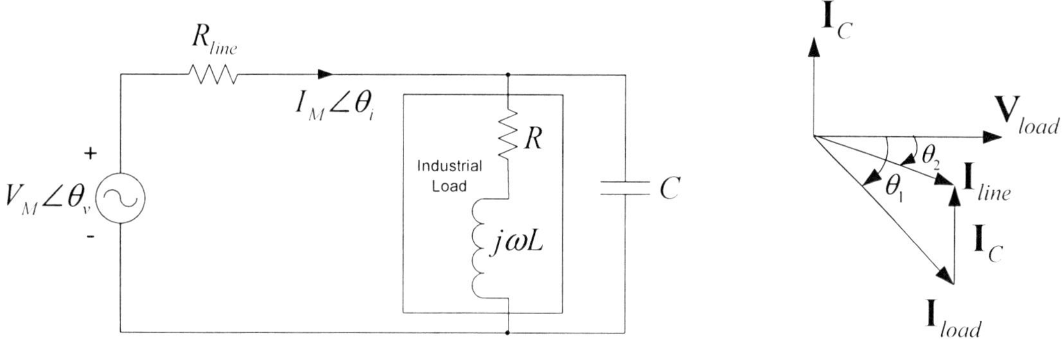

Figure 9.15 Power factor correction

Figure 9.16 New power factor is $\cos\theta_2$

Example 55 – Power Factor Correction

In Fig. 9.17, a 60Hz voltage source $\mathbf{V}_S$ supplies a transmission line with a line resistance of $R_{line} = 0.1\ \Omega$. The voltage across an industrial load is $240\angle 0°$ Vrms and the industrial load is 120 kW with a lagging power factor $PF_{old} = 0.8$.

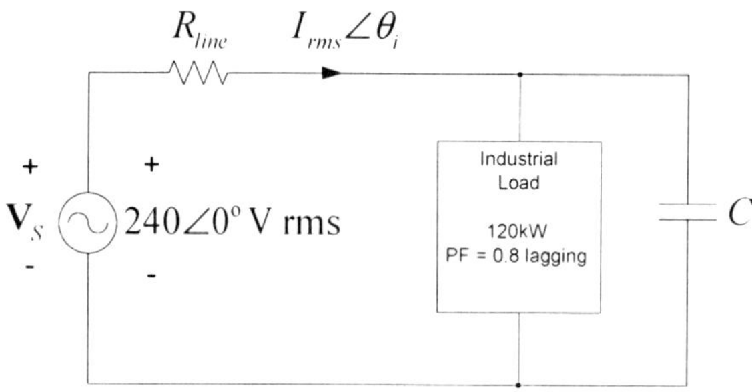

Figure 9.17 Example 55

(a) Calculate the power lost in the line; (b) find the value of capacitance C that will raise the power factor to $PF_{new} = 0.95$; and (c) calculate the power lost in the line after power factor correction.

The following calculations are given in Matlab Example 55.

(a) The power factor angle θ_1 is

$$\theta_1 = \cos^{-1}(PF_{old}) = \cos^{-1}(0.8) = 36.87^\circ \qquad (9.49)$$

From the power triangle shown in Fig. 9.18, we can write

$$Q_1 = 120\,\text{kW}\,\tan(36.87^\circ) = 90.0\ \text{kVAR} \qquad (9.50)$$

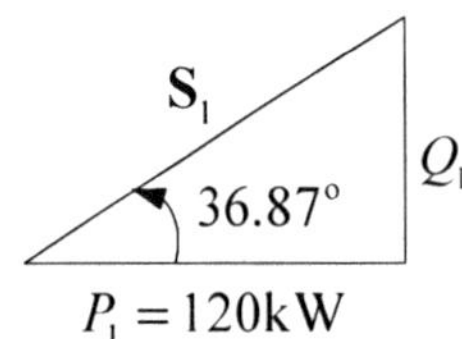

Figure 9.18 Power triangle for PF = 0.8

The value of $\mathbf{S}_1$ in Fig. 9.18 is

$$\mathbf{S}_1 = P_1 + jQ_1 = 120\times10^3 + j90\times10^3 = 150\times10^3\ \angle 36.87^\circ \qquad (9.51)$$

From Eq. (9.43), we can calculate the magnitude of the current I_{rms1} as

$$I_{rms1} = \frac{S_1}{V_{rms}} = \frac{150\times10^3}{240} = 625\ \text{A rms} \qquad (9.52)$$

The real power lost in the line is then

$$P_{line1} = I_{rms1}^2 R_{line} = (625\ \text{A})^2 (0.1\ \Omega) = 39.063\ \text{kW} \qquad (9.53)$$

(b) The new power factor angle θ_2 is

$$\theta_2 = \cos^{-1}(PF_{new}) = \cos^{-1}(0.95) = 18.19^\circ \qquad (9.54)$$

Because we still want to deliver 120kW of real power to the load, the new power triangle is shown in Fig. 9.19. From this triangle we can write

$$Q_2 = 120 \text{ kW} \tan\left(18.19^\circ\right) = 39.442 \text{ kVAR} \qquad (9.55)$$

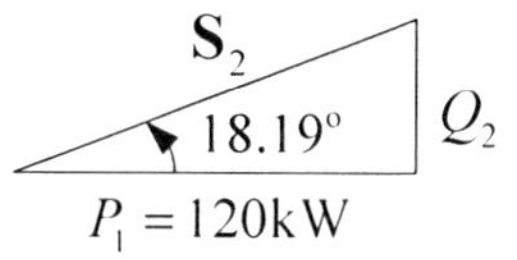

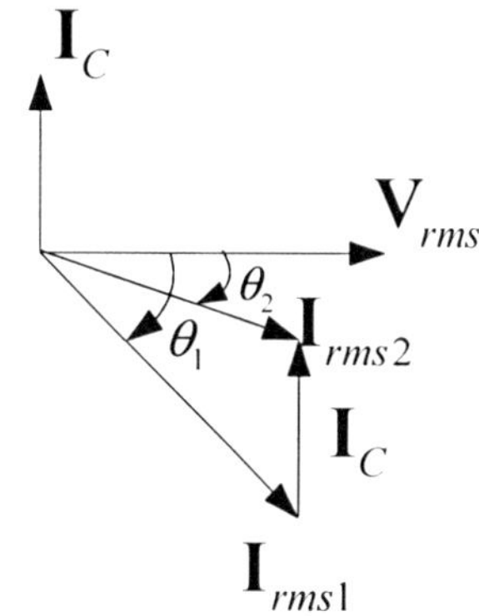

Figure 9.19 Power triangle for PF = 0.95

The new value of $\mathbf{S}_2$ in Fig. 9.19 is

$$\mathbf{S}_2 = P_1 + jQ_2 = 120\times10^3 + j39.442\times10^3 = 126.32\times10^3 \angle 18.19^\circ \quad (9.56)$$

From Eq. (9.43) we can calculate the magnitude of the current I_{rms2} as

$$I_{rms2} = \frac{S_2}{V_{rms}} = \frac{126.32\times10^3}{240} = 526.32 \text{ A rms} \qquad (9.57)$$

The phasor diagram corresponding to Fig. 9.16 is shown in Fig. 9.20. The original load current $\mathbf{I}_{rms1}$ is the same. The new line current $\mathbf{I}_{rms2}$ is the sum of $\mathbf{I}_{rms1}$ and the current through the capacitor $\mathbf{I}_C$.

Figure 9.20 Power triangle for PF = 0.95

We can write the phasor current $\mathbf{I}_{rms1}$ as

$$\mathbf{I}_{rms1} = I_{rms1}e^{-j\theta_1} = 625e^{-j36.87^\circ} \text{ A rms} \qquad (9.58)$$

and the phasor current $\mathbf{I}_{rms2}$ as

$$\mathbf{I}_{rms2} = I_{rms2}e^{-j\theta_2} = 526.32e^{-j18.19^\circ} \tag{9.59}$$

The current $\mathbf{I}_C$ is then

$$\mathbf{I}_C = \mathbf{I}_{rms2} - \mathbf{I}_{rms1} = 210.66e^{j\pi/2} \tag{9.60}$$

This capacitor current is purely reactive (imaginary) and its magnitude will be given by

$$I_C = \frac{V_{rms}}{1/\omega C} = \omega V_{rms}C = 210.66 \tag{9.61}$$

from which the capacitance can be found to be

$$C = \frac{I_C}{\omega V_{rms}} = \frac{210.66}{2\pi 60(240)} = 0.0023 \ \mathrm{F} = 2300 \ \mu\mathrm{F}$$

Matlab Example 55

```
>> Vrms = 220;
>> Rline = 0.1;
>> P1 = 120*10^3;
>> PFold = 0.8;
>> PFnew = 0.95;
>> f = 60;
>> Vrms = 240;
>> theta1 = acos(PFold)
theta1 =
    0.6435

>> theta1d = theta1*180/pi
theta1d =
    36.8699

>> Q1 = P1*tan(theta1)
Q1 =
   9.0000e+004

>> S1 = P1+j*Q1
S1 =
   1.2000e+005 +9.0000e+004i

>> Irms1 = abs(S1)/Vrms
Irms1 =
    625
```

Matlab Example 55 (cont.)

```
>> Pline1 = Irms1^2*Rline
Pline1 =
  3.9063e+004

>> theta2 = acos(PFnew)
theta2 =
    0.3176

>> theta2d = theta2*180/pi
theta2d =
   18.1949

>> Q2 = P1*tan(theta2)
Q2 =
  3.9442e+004

>> S2 = P1+j*Q2
S2 =
  1.2000e+005 +3.9442e+004i

>> Irms2 = abs(S2)/Vrms
Irms2 =
  526.3158

>> I1 = Irms1*exp(-j*theta1)
I1 =
  5.0000e+002 -3.7500e+002i

>> I2 = Irms2*exp(-j*theta2)
I2 =
  5.0000e+002 -1.6434e+002i

>> Ic = I2 - I1
Ic =
  5.6843e-014 +2.1066e+002i

>> Ic = imag(Ic)
Ic =
  210.6579

>> C = Ic/(w*Vrms)
C =
    0.0023

>> Pline2 = Irms2^2*Rline
Pline2 =
  2.7701e+004

>> deltaPline = Pline1 - Pline2
deltaPline =
  1.1362e+004
```

(c) After the power factor correction, the power lost in the line will be

$$P_{line2} = I_{rms2}^2 R_{line} = (526.32 \text{ A})^2 (0.1 \ \Omega) = 27.701 \text{ kW} \qquad (9.62)$$

Therefore the power savings to the generating plant will be

$$\Delta P_{line} = P_{line1} - P_{line2} = 39.063 \text{ kW} - 27.701 \text{ kW} = 11.362 \text{ kW} \qquad (9.63)$$

Problems

9.1 For the circuit shown below, plot the currents $i_R(t)$, $i_L(t)$, $i_C(t)$, and the instantaneous power $p(t)$ as a function of time for two cycles of the input current. Let $R = 600 \ \Omega$, $C = 2.653$ nF, $L = 9.5$ mH, $V_M = 1$ V, and $f = 10^5$ Hz.

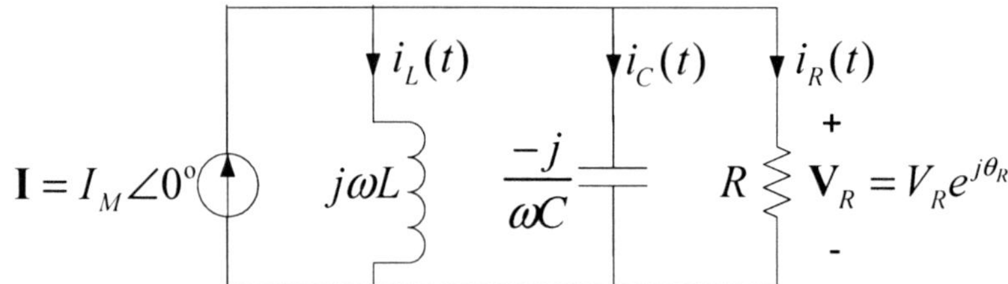

Chapter 10

DC Motors

Electric motors are complex devices that require an entire course of their own. There are many different kinds of electric motors including DC motors, stepper motors, servo motors, AC motors, and brushless DC motors. In this chapter, we will discuss the basic operation of only the simplest kind of DC motor. We will introduce a simple equivalent circuit for a DC motor that will allow you to calculate some basic properties of a DC motor.

10.1 Operation of a DC Motor

In Chapter 1, we saw that Faraday was able to generate rotational motion with his electromagnetic rotator shown in Fig. 1.3. We also learned that Thomas Davenport received the first U. S. patent for a DC electric motor in 1837. To understand how a DC motor works, consider a loop of wire carrying a current i in a magnetic field $\mathbf{B}$ as shown in Fig. 10.1a.

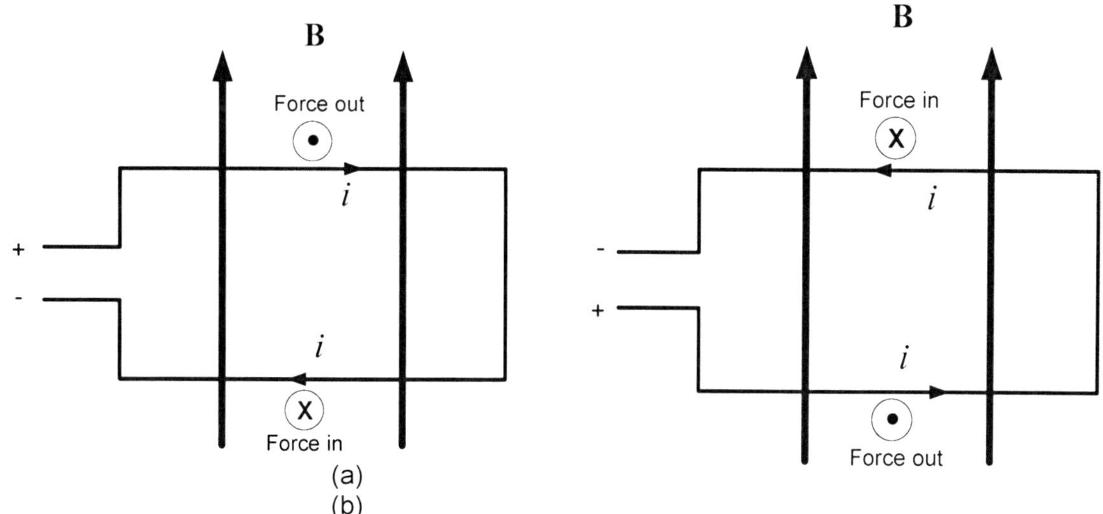

Figure 10.1 Forces on a current loop in a magnetic field

Recall from Eq. (1.5) in Chapter 1 that Ampère discovered that the force per unit length of a current-carrying conductor in a magnetic field is given by

$$\frac{\Delta \mathbf{F}}{\Delta L} = \mathbf{i} \times \mathbf{B} \qquad\qquad (10.1)$$

Applying this cross product to the current and magnetic field in Fig. 10.1a, we see that the force on the top horizontal wire will be out of the page while the force on the bottom horizontal wire will be into of the page. There will be no force on the sections of wire that are parallel to the magnetic field. Thus, this coil of wire will tend to rotate about a horizontal axis along the centerline of the coil, that is, it will rotate counter-clockwise when looking at the coil from the right end.

Now consider the situation after the coil has rotated 180° as shown in Fig. 10.1b. The + terminal is now at the bottom so the current flows to the right in the bottom horizontal wire and flows to the left in the top horizontal wire. In this case, the forces on the top and bottom horizontal wires are just the opposite of those in Fig. 10.1a and the coil will tend to rotate clockwise when looking at the coil from the right end. Thus, we would expect the coil to just rock back and forth and not rotate like an electric motor should.

To solve this problem, we need to have the current in the top horizontal wire always flowing to the right and have the current in the bottom horizontal wire always flowing to the left. To do this, we need to have the + terminal always stay on top. A solution is to use a *commutator* as shown in Fig. 11.2. William Sturgeon, an English physicist and inventor, first demonstated an electric motor using a commutator in 1832. Thomas Davenport also used a commutator in his electric motor.

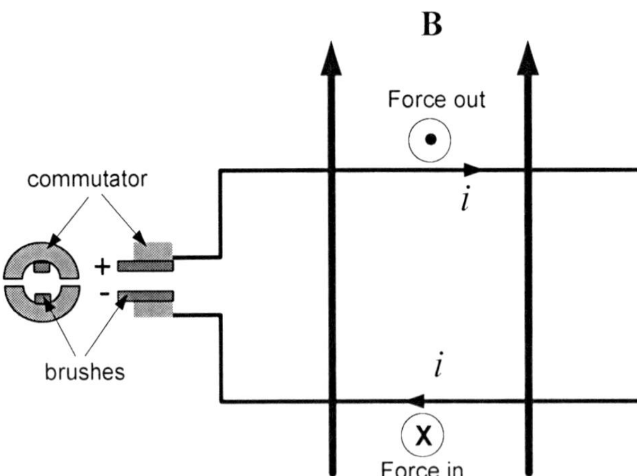

Figure 10.2 A commutator always keeps the current in the top wire flowing to the right

In Fig. 10.2, the brushes are connected to a DC power supply and do not move. The commutator is a split ring with each half connected to one end of the coil. An electrical connection is made between the brushes and the commutator as the commutator slides over the brushes as the coil rotates. The commutator acts like a switch that switches the polarity of the coil every half cycle. Thus, the current in the top horizontal wire in Fig. 10.2 is always flowing to the right and the force on that wire is always out of the page. Thus, the coil will continue to rotate in a counter-clockwise direction when looking at the coil from the right end.

For a constant current, the two forces shown in Fig. 10.2 will produce a constant torque on the coil, which by Newton's law should cause the coil to have a constant angular acceleration. This means that the coil will spin faster and faster without limit! We know that this can't be so, but why? If you're thinking that the friction will somehow slow it down, that won't work. We could always increase the current so that the net torque on the coil is positive, which would continue to accelerate the coil. The coil doesn't speed up forever for a much more fundamental reason, one that we have already seen in both Chapters 1 and 2. Remember Faraday's law of electromagnetic induction (as expressed in Maxwell's 3rd equation in Fig. 1.14) that states that the rate of change of magnetic flux density through the area of a coil induces an electromotive force (*emf*), or voltage, across the terminals of the coil. We used this law in Chapter 2 where we saw in Fig. 2.23 that an alternating current through the coil will produce a changing magnetic flux density, which by Faraday's law will induce a voltage across the coil. This is the source of the voltage across an inductor. In that case, it was the AC current that produced the changing magnetic field. But in Fig. 10.2, the magnetic field is constant. However, when the coil rotates, the number of magnetic flux lines that pierce the area of the coil will vary. At the instant of time shown in Fig. 10.2, no magnetic flux lines go through the area of the coil but, when the coil has rotated $90°$, a maximum number of magnetic flux lines will go through the area of the coil. Thus, the number of magnetic flux lines going through the area of the coil will vary with time and, by Faraday's law, this will induce a voltage at the terminals of the coil and, by Lenz's law, the sign of the voltage will be such as to produce a current in the coil that is opposite to the current producing the torque on the coil. This induced voltage, called a *back emf*, will keep the coil from speeding up forever.

Back Emf

Let's see of we can derive an expression for the back emf that gets induced in the rotating coil. In Fig. 10.3, we show the coil after it has rotated through an angle θ from the x-axis. To find the total magnetic flux ϕ going through the coil, we multiply the magnetic flux density **B** by the component of the coil area that is normal to **B**. This area A is lw where l is the length of the coil and $w = 2r\cos\theta$ is the projection of the coil width onto the x-axis as shown in Fig. 10.3. Thus, the total flux $\phi(t)$ going through the coil at any instant of time is given by

$$\phi = \mathbf{B}A = \mathbf{B}2rl\cos\theta \qquad (10.2)$$

Now, by Faraday's law, the induced emf is given by

$$\text{emf} = -\frac{\partial \phi}{\partial t} = -\frac{\partial}{\partial t}\left(\mathbf{B}2rl\cos\theta\right) \qquad (10.3)$$

or

$$\text{emf} = e_{ab} = \mathbf{B}2rl\sin\theta\,\frac{\partial\theta}{\partial t} = 2\mathbf{B}rl\omega\sin\theta \qquad (10.4)$$

where

$$\omega = \frac{d\theta}{dt} \qquad (10.5)$$

is the angular velocity of the coil.

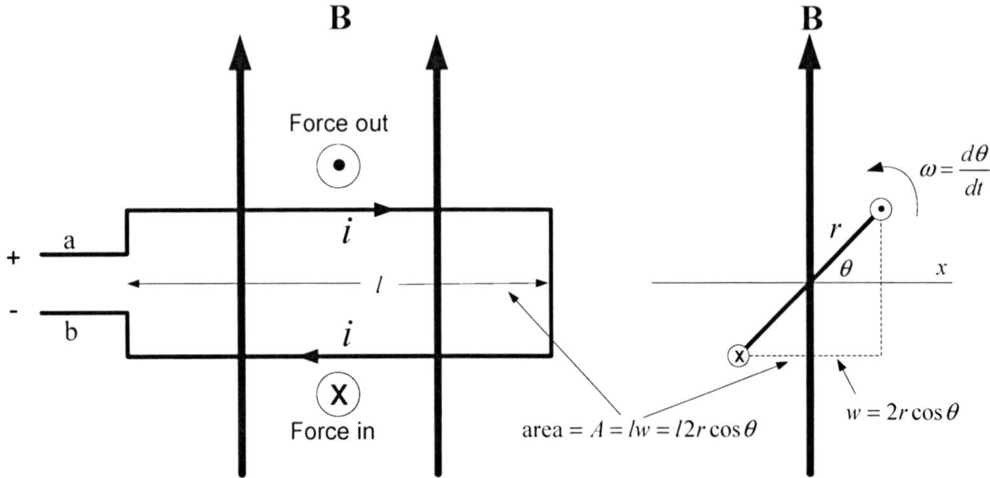

Figure 10.3 Deriving an expression for the back emf

We therefore see that the back emf, e_{ab}, varies sinusoidally as the coil rotates as shown in Fig. 10.4 and the magnitude of e_{ab} is proportional to the angular velocity ω.

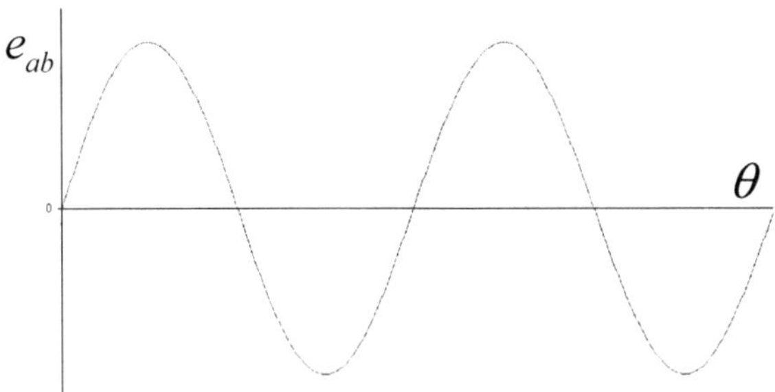

Figure 10.4 A plot of $e_{ab} = 2\mathbf{B}rl\omega\sin\theta$

Now, if we were to add a commutator of the type shown in Fig. 10.2, the value of e_{ab} would always be positive as shown in Fig. 10.5. Finally, if we were to add three more coils offset by $45°$ and add the back emfs together, we would obtain an output that would look like Fig. 10.6.

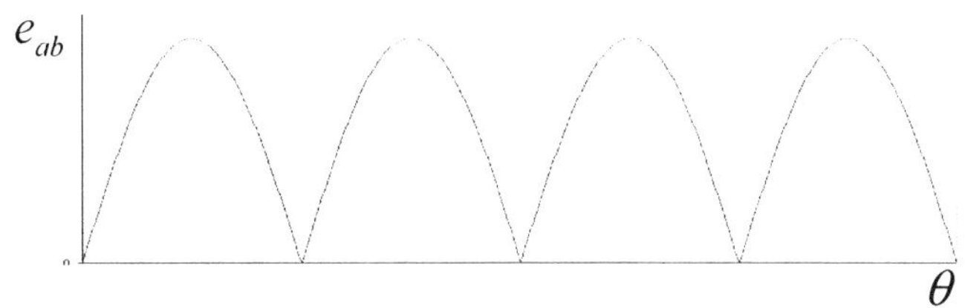

Figure 10.5 A plot of $e_{ab} = 2\mathbf{B}rl\omega|\sin\theta|$

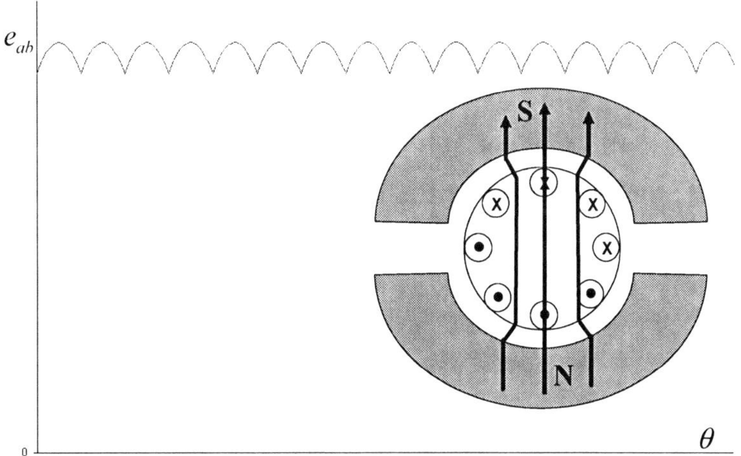

Figure 10.6 An armature with four coil loops

We therefore see that the total back emf of a motor with many coils will be a DC voltage that is proportional to the angular velocity ω. We write this induced back emf as

$$E_a = K_a\phi\omega \qquad (10.6)$$

where $K_a\phi$ is a constant that depends on the motor's armature geometry and the magnet's flux density. We can then summarize our discussion of the basic operation of a DC motor by drawing the equivalent circuit shown in Fig. 10.7. In this circuit, V_t is the terminal voltage to the motor, R_a is the resistance of the armature coils, I_a is the armature current, and E_a is the induced back emf. In the next section, we will use this simple equivalent circuit to derive some important equations relating to the operation of a DC motor.

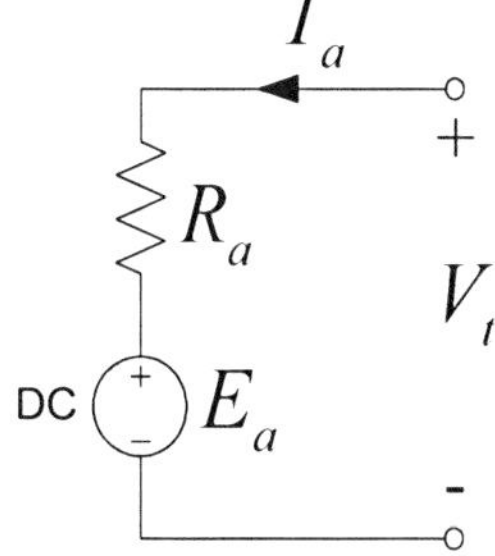

Figure 10.7 Equivalent circuit for a simple DC motor

10.2 Motor Equations

Applying KVL to the equivalent circuit for a DC motor shown in Fig. 10.7, we can write

$$V_t = E_a + I_a R_a \tag{10.7}$$

or, solving for I_a,

$$I_a = (V_t - E_a)/R_a \tag{10.8}$$

Substituting Eq. (10.6) into Eq. (10.7), we obtain

$$I_a = (V_t - K_a\phi\omega)/R_a \tag{10.9}$$

Eq. (10.9) tells us a lot about how the motor works. When the voltage V_t is first applied to the motor terminals, the motor is not rotating and thus $\omega = 0$ and the armature current will be $I_a = V_t/R_a$. This armature current will cause the motor to rotate, which will increase the back emf term $K_a\phi\omega$. When this term becomes equal to V_t, the armature current will become zero and no further increase in motor speed is possible. Therefore, the maximum motor speed will be

$$\omega_{max} = \frac{V_t}{K_a\phi} \tag{10.10}$$

Suppose we were to grab a hold of the motor shaft and mechanically rotate it faster than ω_{max}. Then the armature current in Eq. (10.9) will become negative and we will be providing current to the rest of the circuit. We have just turned the motor into a generator. In fact, if you just take a DC motor and mechanically rotate its shaft, it will generate a back emf voltage that will produce a generated voltage at the terminals as shown in Fig. 10.8. Note that the only difference between the equivalent circuit of a generator and the equivalent circuit of a DC motor in Fig. 10.7 is the direction of the armature current. From Fig. 10.8, we see that the equation for the generated voltage is given by

$$V_t = E_a - I_a R_a \tag{10.11}$$

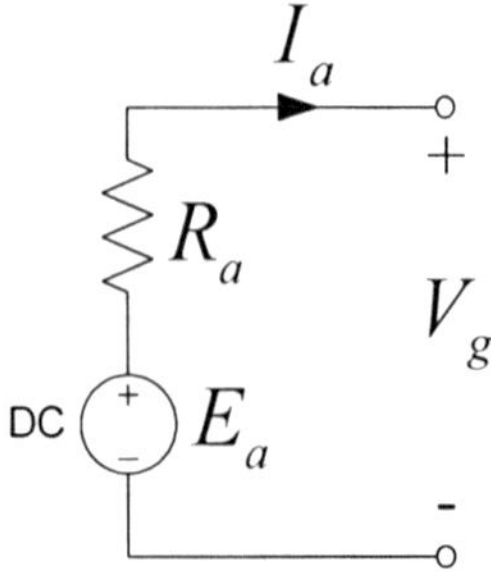

Figure 10.8 Equivalent circuit for a generator

Power and Torque

The electrical power delivered to the rotating motor is equal to $E_a I_a$ and is converted to mechanical power, called *developed power* P_d that is equal to the *developed torque* τ_d times the angular speed ω in rad/s. Thus, we can write

$$P_d = E_a I_a = \tau_d \omega \tag{10.12}$$

Substituting Eq. (10.6) into Eq. (10.12) and solving for τ_d, we obtain

$$\tau_d = K_a \phi I_a \tag{10.13}$$

or

$$I_a = \frac{1}{K_a \phi} \tau_d \tag{10.14}$$

Thus, the developed torque is proportional to the armature current and vice versa.
Substituting Eq. (10.9) into Eq. (10.13), we obtain

$$\tau_d = \frac{V_t K_a \phi}{R_a} - \frac{\left(K_a \phi\right)^2}{R_a} \omega \tag{10.15}$$

or

$$\omega = \frac{V_t}{K_a \phi} - \frac{R_a}{\left(K_a \phi\right)^2} \tau_d \tag{10.16}$$

Thus, as the motor speed ω increases, the developed torque decreases linearly, or alternatively, as the torque increases, the motor speed decreases.
If we substitute Eq. (10.16) into Eq. (10.12), we can write the developed power P_d as

$$P_d = \frac{V_t}{K_a \phi} \tau_d - \frac{R_a}{\left(K_a \phi\right)^2} \tau_d^2 \tag{10.17}$$

Thus, the developed power is a quadratic function of the torque. The maximum power will occur when

$$\frac{dP_d}{d\tau_d} = \frac{V_t}{K_a \phi} - \frac{2R_a}{\left(K_a \phi\right)^2} \tau_d = 0 \tag{10.18}$$

or when

$$\tau_d = \tau_{d\,\text{max}} = \frac{V_t K_a \phi}{2R_a} \tag{10.19}$$

Substituting this value for $\tau_{d\,\text{max}}$ back in Eq. (10.17) gives a maximum power of

$$P_{d\,\max} = \frac{V_t^2}{4R_a} \tag{10.20}$$

Example 56 – Motor Speed

A DC motor has a terminal voltage of 5 V, an armature resistance of 4 Ω, and a value of $K_a\phi = 6.4 \times 10^{-3}$ volt-sec. Calculate the stall torque and the no-load speed. Plot the motor speed in RPM vs. torque.

The stall torque occurs when $\omega = 0$ in Eq. (10.15). Therefore,

$$\tau_{stall} = \frac{V_t K_a\phi}{R_a} = 0.008\text{N-m} \tag{10.21}$$

The no-load speed occurs when $\tau_d = 0$ in Eq. (10.16). Therefore,

$$\omega_{no-load} = \frac{V_t}{K_a\phi} = 7460.4\text{RPM} \tag{10.22}$$

The Matlab function *motorspeed(Vt,Ra,Kphi)* given in Listing 10.1 will solve this problem. The solution is shown in Matlab Example 56 and the resulting plot is shown in Fig. 10.9. Note that the speed of the motor decreases linearly with increasing torque.

Listing 10.1 motorspeed.m

```
function motorspeed(Vt,Ra,Kphi)
% find stall_torque in N-m and no_load_speed in RPM
% plot speed as a function of torque
% Vt = terminal voltage in volts
% Ra = armature current in amperes
% Kphi = motor constant in volt-sec
% motorspeed(Vt,Ra,Kphi)
stall_torque = Vt*Kphi/Ra
no_load_speed = (Vt/Kphi)*(60/(2*pi))
tau = linspace(0,stall_torque,100);
w = no_load_speed - (no_load_speed/stall_torque).*tau;
plot(tau,w)
```

Matlab Example 56

```
>> help motorspeed
   find stall_torque in N-m and no_load_speed in RPM
   plot speed as a function of torque
   Vt = terminal voltage in volts
   Ra = armature current in amperes
   Kphi = motor constant in volt-sec
   motorspeed(Vt,Ra,Kphi)

>> Vt = 5;
>> Ra = 4;
>> Kphi = 6.4*10^-3;
>> motorspeed(Vt,Ra,Kphi)
stall_torque =
     0.0080

no_load_speed =
   7.4604e+003
```

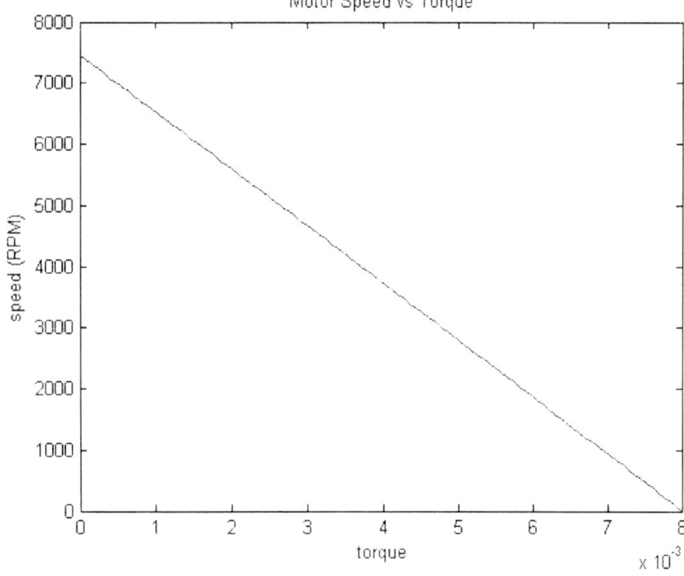

Figure 10.9 Plot resulting from Matlab Example 41

Example 57 – Armature Current and Power

Using the same DC motor as in Example 56 with a terminal voltage of 5 V, an armature resistance of 4 Ω, and a value of $K_a \phi = 6.4 \times 10^{-3}$ volt-sec, calculate the stall current, the maximum power, and the current and torque at maximum power. Plot the current and power vs. torque.

The Matlab function *IP(Vt,Ra,Kphi)* given in Listing 10.2 will solve this problem. The solution is shown in Matlab Example 57 and the resulting plot is shown in Fig. 10.10.

Listing 10.2 IP.m

```
function IP(Vt,Ra,Kphi)
% find stall_current in amperes
% find the maximum power in watts
% find the current and torque (in N-m) at maximum power
% plot current and power as a function of torque
% Vt = terminal voltage in volts
% Ra = armature current in amperes
% Kphi = motor constant in volt-sec
% IP(Vt,Ra,Kphi)
stall_torque = Vt*Kphi/Ra;
stall_current = Vt/Ra
Pmax = Vt^2/(4*Ra)
tau_max = Vt*Kphi/(2*Ra)
Imax = Vt/(2*Ra)
tau = linspace(0,stall_torque,100);
i = tau./Kphi;
Pd = (Vt/Kphi).*tau - (Ra/Kphi^2).*tau.^2;
curves = [i;Pd];
plot(tau,curves)
```

Matlab Example 57

```
>> help IP
  find stall_current in amperes
  find the maximum power in watts
  find the current and torque (in N-m) at maximum power
  plot current and power as a function of torque
  Vt = terminal voltage in volts
  Ra = armature current in amperes
  Kphi = motor constant in volt-sec
  IP(Vt,Ra,Kphi)

>> Vt = 5;
>> Ra = 4;
>> Kphi = 6.4*10^-3;
>> IP(Vt,Ra,Kphi)

stall_current =
    1.2500

Pmax =
    1.5625

tau_max =
    0.0040

Imax =
    0.6250
```

The stall current can be found by substituting Eq. (10.21) in Eq. (10.14). Therefore,

$$I_{stall} = \frac{V_t}{R_a} = 1.25 \text{ A} \qquad (10.23)$$

The maximum power is given by Eq. (10.20) and is equal to

$$P_{d\max} = \frac{V_t^2}{4R_a} = 1.5625 \text{ W} \qquad (10.24)$$

This maximum power occurs at a value of torque given by Eq. (10.19) and is equal to

$$\tau_{d\max} = \frac{V_t K_a \phi}{2R_a} = 0.004 \text{ N-m} \qquad (10.25)$$

Substituting this value for $\tau_{d\max}$ in Eq. (10.14) gives the current at maximum power as

$$I_{a\max} = \frac{V_t}{2R_a} = 0.625 \text{ A} \qquad (10.26)$$

Note from the plot in Fig. 10.10 that the armature current increases linearly with increasing torque and that the maximum power occurs at one half the stall torque.

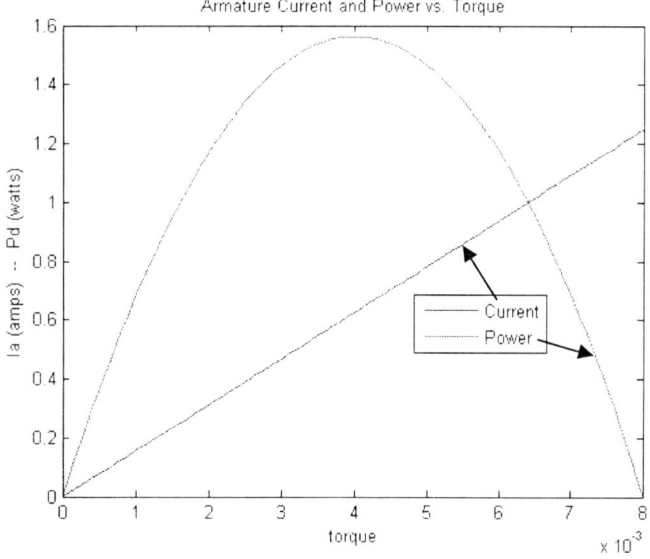

Figure 10.10 Plot resulting from Matlab Example 57

The data sheets of DC motors will normally include graphs similar to Figs. 10.9 and 10.10.

Problems

10.1 Suppose that you have two identical DC motors with armature resistances of 4 Ω and values of $K_a\phi = 6.4\times10^{-3}$ volt-sec as were used in Matlab Examples 56 and 57. If you connect both motor shafts together and apply a terminal voltage $V_t = 5$ V to the first motor, the second motor shaft will rotate at the same angular speed as the first motor. The second motor will then act as a generator and produce a voltage V_g at its terminals. Assume that the first motor's developed torque is such as to produce the maximum power. (a) What is the speed of the first motor in RPM? (b) What is the generated voltage V_g ?

10.2 Describe how you could change the speed of a DC motor.

Appendix A

Addition and Subtraction of Vectors

In this appendix, the basic elements of vector algebra are explored. Vectors are treated as geometric entities represented by directed line segments. The ways that the components of a vector can be written in Matlab will be introduced.

Prerequisite knowledge:
    Basic trigonometry and plane geometry
    Algebra including determinants

A.1 Scalars and Vectors

Different types of physical quantities can be distinguished by the number of pieces of information required to completely specify them. For example, temperature and mass have only a magnitude and thus a single number representing this magnitude is sufficient to completely specify them. Such physical quantities are called *scalars*. Thus a scalar is a physical quantity that is specified by giving only a magnitude or single number.

Certain quantities, however, cannot be completely specified by a magnitude only. These are quantities such as velocity and electric field strength, which in addition to a magnitude also have a certain direction associated with them. Such quantities are called *vectors*. We shall see that in three-dimensional space vectors require three pieces of information to completely specify them. A vector then is a physical quantity that is specified by giving both a magnitude and a direction.

There are other kinds of physical quantities that require more than three pieces of information to completely specify them and are therefore neither scalars nor vectors. Such quantities are called *tensors*.

Since a vector has both a magnitude and a direction, we can represent it by a directed line segment or an arrow in which the length of the line segment is proportional to the magnitude of the vector and the orientation of the line segment specifies the direction of the vector. Thus, in Fig. A.1, the magnitude of the vector **A** is equal to two times the magnitude of the vector **B** and the directions of **A** and **B** differ by 90 degrees.

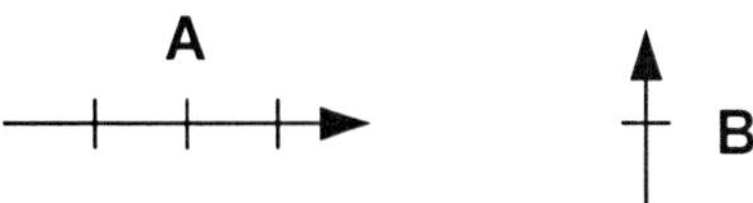

Figure A.1 Vectors have a magnitude and direction

Notice that we have written the vectors **A** and **B** in boldface type to distinguish them from scalars, which are written in italics (*T,m*). In writing vectors by hand, you cannot make a symbol boldface and so some convention must be used to indicate that the symbol refers to a vector quantity. A useful convention is to write a vector quantity as *A* with a wavy line under the symbol. This is useful because, if the material is printed, the typesetter will automatically make such symbols boldface. Another common convention is to use an arrow over the symbol as in $\vec{A}$. .

A.2 Addition of Vectors

Since vectors are specified by giving only a magnitude and a direction, we can relocate the vectors **A** and **B** in Fig. A.1 provided we preserve their original orientation. The vectors are said to be invariant to translation. If we think of **A** and **B** as two successive displacement vectors that describe a person walking four units east and then two units north,, it is clear that the resultant vector **R** = **A** + **B** can be found by placing **B** such that its tail is at the same point as the head of **A**. The resultant vector **R** then has its tail at the tail of **A** and its head at the head of **B** as shown in Fig. A.2.

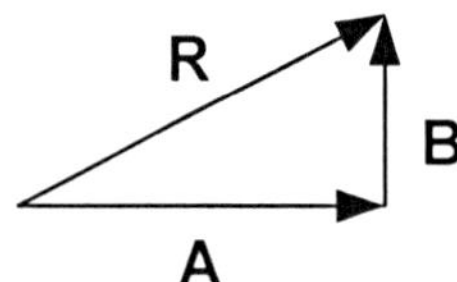

Figure A.2 The resultant vector **R** = **A** + **B**

Now move **B** in Fig. A.2 so that its tail coincides with the tail of **A**. The resultant **R** is seen to lie along the diagonal of the parallelogram formed by **A** and **B**, with the tails of all three vectors coinciding. Finally, shift **A** in Fig. A.3 such that its tail coincides with the head of the shifted **B** as shown in Fig. A.4. It is now obvious from Fig. A.4 that **R** = **A** + **B** = **B** + **A**, showing that the *commutative law* holds for vector addition.

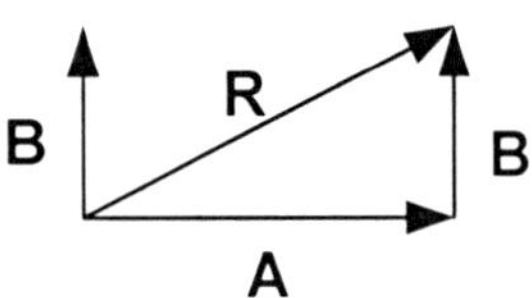

Figure A.3 Shift **B**

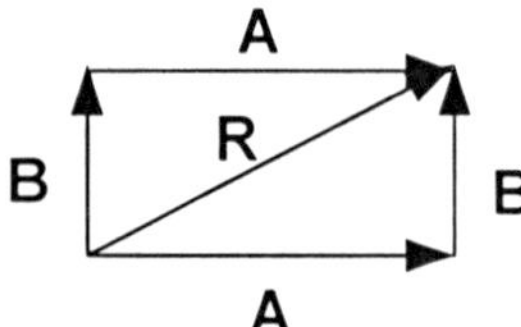

Figure A.4 Shift **A**

The sum of more than two vectors can be found by continuing to place the tail of succeeding vectors at the head of the preceding vector, as shown in Fig. A.5. The resultant vector **D** = **A** + **B** + **C** is shown in Fig. A.6.

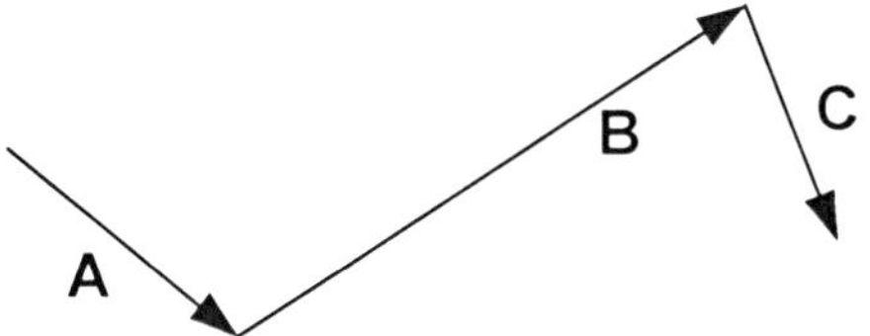

Figure A.3 **A** + **B** + **C**

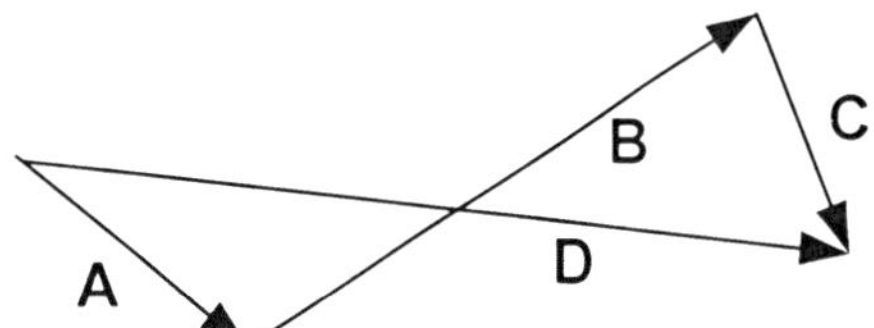

Figure A.4 **D** = **A** + **B** + **C**

Draw the vector sum (**A** + **B**) in Fig. A.4. The result is shown in Fig. A.5. Finally, draw the vector sum (**B** + **C**) in Fig. A.5. The result is shown in Fig. A.6. It is now clear from Fig. A.6 that   **D** = **A** + **B** + **C** = (**A** + **B**) + **C** = **A** + (**B** + **C**), showing that the associative law holds for vector addition.

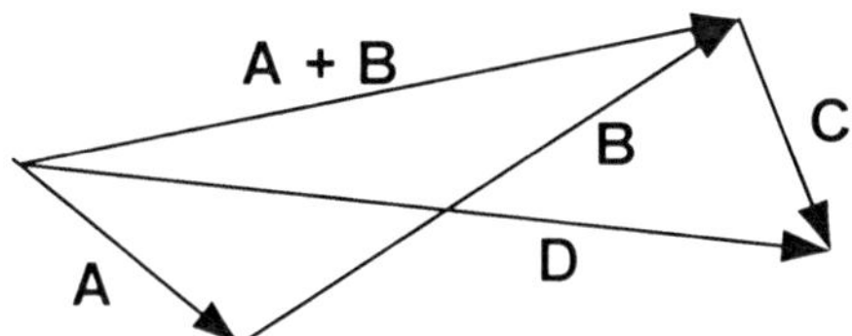

Figure A.5 **D** = (**A** + **B**) + **C**

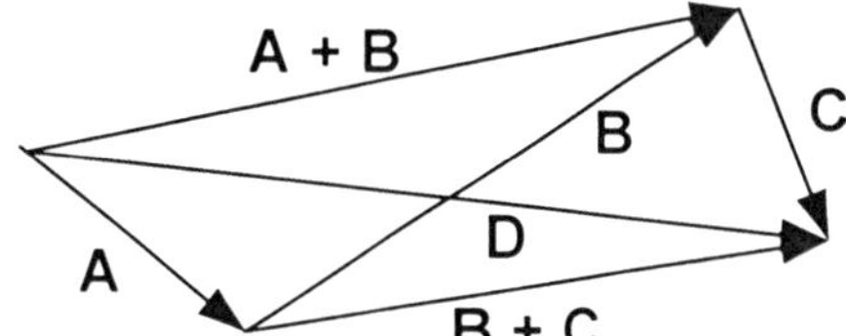

Figure A.6 **D** = **A** + (**B** + **C**)

A.3 Subtraction of Vectors

Two vectors **A** and **B** are shown Fig. A.7. The vector -**A** is a vector with the same magnitude as **A** but with the opposite direction. Draw -**A** in Fig. A.7. The result is shown in Fig. A.8.

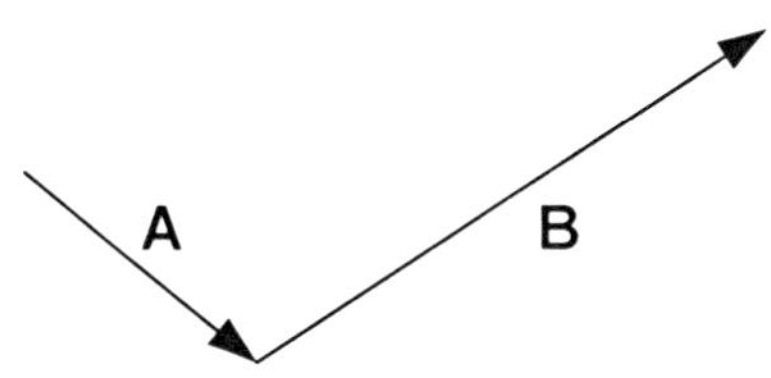

Figure A.7 Vectors **A** and **B**

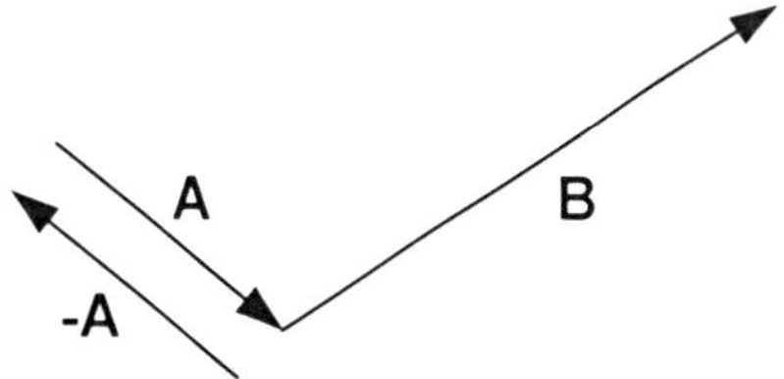

Figure A1.8 The vector -**A**

Draw the resultant vector **R** = **A** + **B** in Fig. A.8. Since **R** = **A** + **B**, verify that **R** - **A** = **B** by showing that **R** + (-**A**) = **B** in Frame A.8. The result is shown in Fig. A.9.

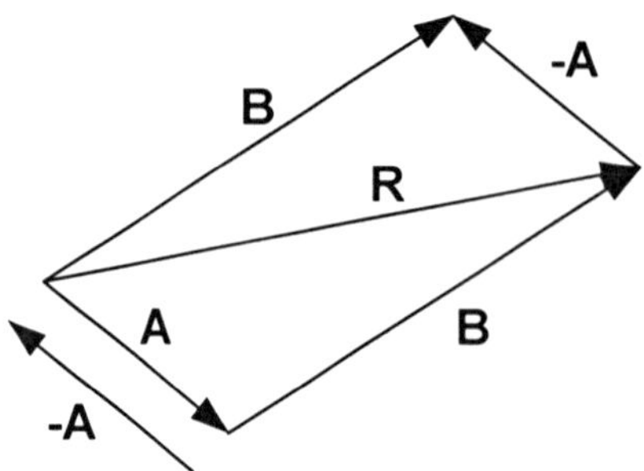

Figure A.9 **R - A = B**

Fig. A.10 shows two vectors **A** and **B**. Draw the resultant vector **R** = **A** + **B** and the difference vector **D** = **A** - **B**. Note that the difference vector **D** can be drawn by connecting the head of **A** with the head of **B** and locating the head of **D** at the head of **A** as shown in Fig. A.11.

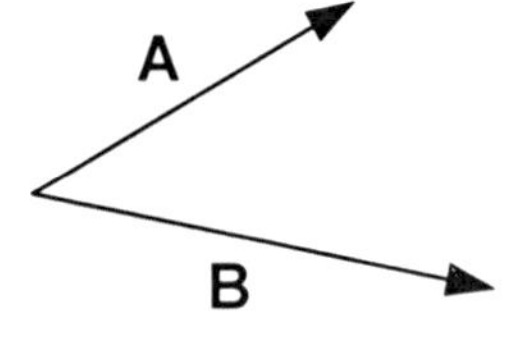

Figure A.10 Vectors **A** and **B**

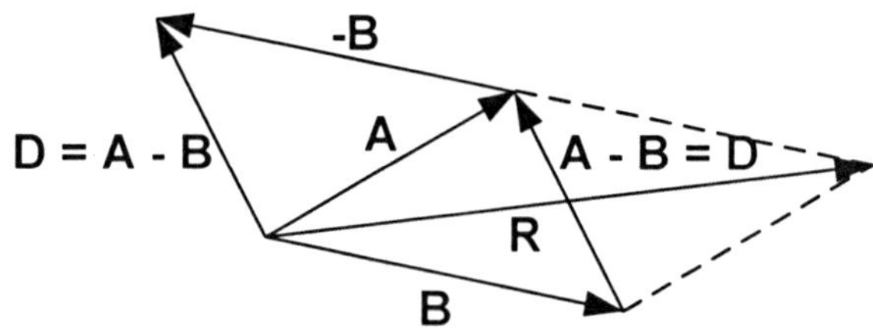

Figure A.11 **D = A - B**

A.4 Unit Vectors and Coordinate Systems

If a vector **A** is multiplied by a scalar m, the resulting product m**A** is a vector whose magnitude is equal to $|m|$ times the magnitude of **A**. The direction of m**A** is the same as that of **A** if m is positive and opposite to that of **A** if m is negative. If λ is a vector having a magnitude of unity, then $m\lambda$ is a vector whose magnitude is $|m|$.

The magnitude of the vector **A** is written as $|$**A**$|$ = A. In Fig. A.12, the unit vector λ_A, which has a magnitude of unity, is in the same direction as **A**. We can therefore write the vector **A** as the magnitude of **A** multiplied by the unit vector λ_A. That is, **A** = $A\lambda_A$. The unit vector λ_A in the direction of **A** can then be written as $\lambda_A = \dfrac{\mathbf{A}}{A}$.

Figure A.12 $\mathbf{A} = A\lambda_A$

Fig. A.13 shows $\mathbf{A}$ to be the vector sum of $\mathbf{A}_x$ and $\mathbf{A}_y$. That is, $\mathbf{A} = \mathbf{A}_x + \mathbf{A}_y$. The vectors $\mathbf{A}_x$ and $\mathbf{A}_y$ lie along the x and y axes; therefore, we say that the vector $\mathbf{A}$ has been resolved into its x and y components.

The unit vectors $\mathbf{i}$ and $\mathbf{j}$ are directed along the x and y axes as shown in Fig. A.13. Using the technique of Fig. A.12, we can therefore write $\mathbf{A}_x = A_x\mathbf{i}$ and $\mathbf{A}_y = A_y\mathbf{j}$. We can then write $\mathbf{A}$ in terms of the unit vectors as the vector sum $\mathbf{A} = A_x\mathbf{i} + A_y\mathbf{j}$. From Fig. A.13, we see that the magnitudes are related by

$$A_x = A \cos \theta$$
$$A_y = A \sin \theta$$

from which the ratio

$$\frac{A_y}{A_x} = \frac{\sin \theta}{\cos \theta} = \tan \theta$$

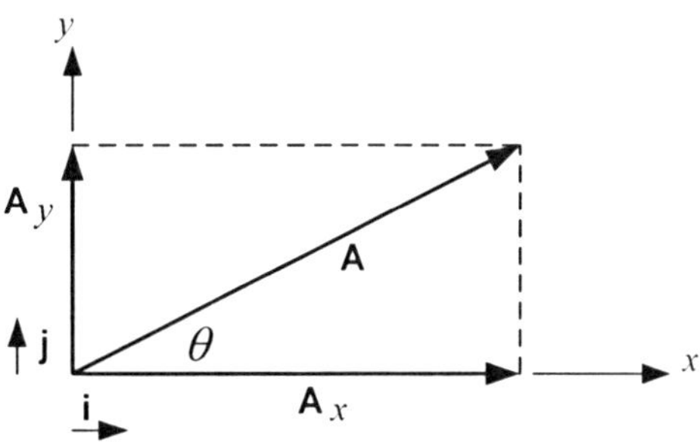

Figure A.13 $\mathbf{A} = A_x\mathbf{i} + A_y\mathbf{j}$

Square A_x and A_y and add the results to obtain $A_x^2 + A_y^2 = A^2\left(\cos^2 \theta + \sin^2 \theta\right) = A^2$ from which $A = \sqrt{A_x^2 + A_y^2}$.

A.5 Addition of Vectors by Components

To illustrate the addition of vectors by components, consider the vector sum $\mathbf{R} = \mathbf{A} + \mathbf{B}$ shown in Fig. A.14. By resolving $\mathbf{A}$ and $\mathbf{B}$ into x and y components, we can write

$$\mathbf{R} = \mathbf{A} + \mathbf{B}$$
$$= A_x\mathbf{i} + A_y\mathbf{j} + B_x\mathbf{i} + B_y\mathbf{j}$$

from which

$$\mathbf{R} = (A_x + B_x)\mathbf{i} + (A_y + B_y)\mathbf{j}$$

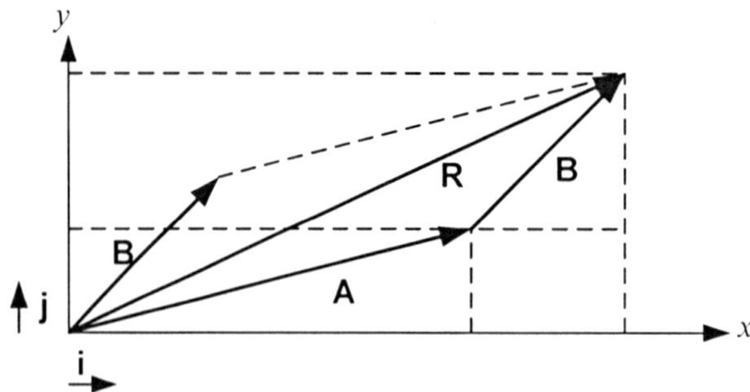

Figure A.14 $\mathbf{R} = (A_x + B_x)\mathbf{i} + (A_y + B_y)\mathbf{j}$

But, in Fig. A.14, $\mathbf{R}$ can be written as $\mathbf{R} = R_x\mathbf{i} + R_y\mathbf{j}$. Therefore,

$$R_x = A_x + B_x$$

$$R_y = A_y + B_y$$

These results are summarized in Fig. A.15.

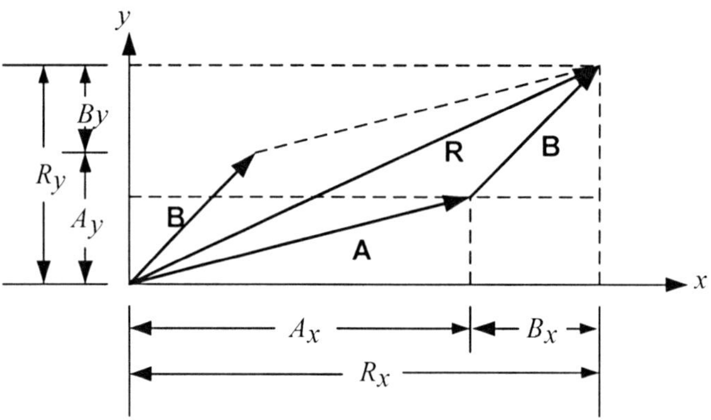

Figure A.15 $\mathbf{R} = R_x\mathbf{i} + R_y\mathbf{j} = (A_x + B_x)\mathbf{i} + (A_y + B_y)\mathbf{j}$

A.6 3-Dimensional Vectors

The results above can readily be extended to three dimensions. From Fig. A.16, the vector $\mathbf{A}$ is the vector sum $\mathbf{A} = \mathbf{A}_x + \mathbf{A}_y + \mathbf{A}_z$ or, in terms of the unit vectors $\mathbf{i}$, $\mathbf{j}$, and $\mathbf{k}$,

$$\mathbf{A} = A_x\mathbf{i} + A_y\mathbf{j} + A_z\mathbf{k}$$

The magnitudes associated with this vector sum are shown in Fig. 1 17. In terms of θ and ϕ, we can write

$$A_x = R \cos \phi$$
$$A_y = R \sin \phi$$
$$A_z = A \cos \theta$$

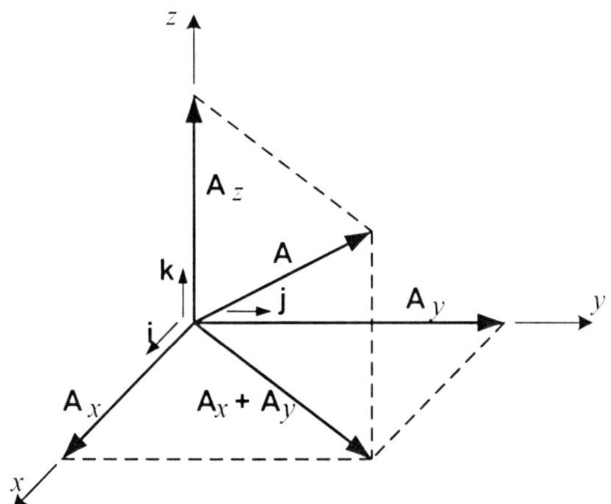

Figure A.16 $\mathbf{A} = A_x\mathbf{i} + A_y\mathbf{j} + A_z\mathbf{k}$

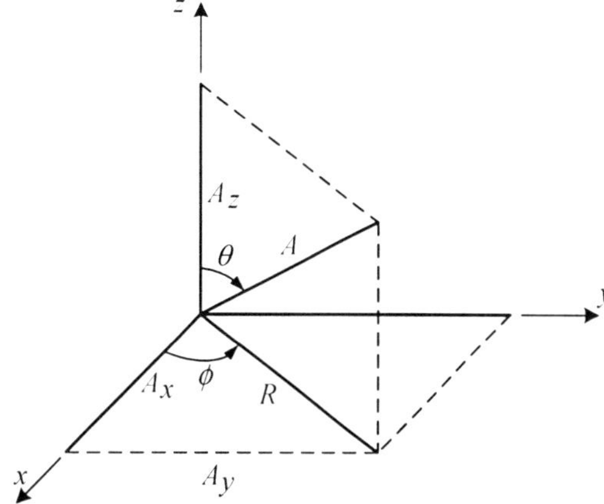

Figure A.17 Magnitudes of the components of $\mathbf{A}$

Since $R = A \sin \theta$, the components in Fig. A.17 can be written as

$$A_x = A \sin \theta \cos \phi$$
$$A_y = A \sin \theta \sin \phi$$
$$A_z = A \cos \theta$$

Square A_x, A_y, and A_z and add the results to obtain

$$A_x^2 + A_y^2 + A_z^2 = A^2 \left[\sin^2 \theta \left(\cos^2 \phi + \sin^2 \phi \right) + \cos^2 \theta \right]$$
$$= A^2 \left[\sin^2 \theta + \cos^2 \theta \right]$$
$$= A^2$$

A.7 Direction Cosines

An alternate way of describing the vector $\mathbf{A}$ in three dimensions is by projecting the vector directly onto the x, y, and z coordinates through the angles α, β, and γ, respectively, as shown in Fig. A.18. Thus,

$$A_x = A \cos \alpha$$
$$A_y = A \cos \beta$$
$$A_z = A \cos \gamma$$

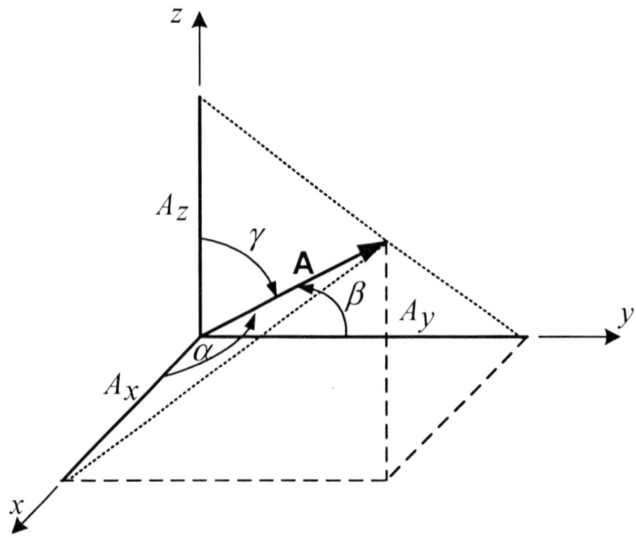

Figure A.18 Definition of direction cosines

The cosines of the angles α, β, and γ in Fig. A.18 are called the *direction cosines* and are designated by *l*, *m*, and *n*, respectively. Thus, in terms of A, A_x, A_y, and A_z

$$l = \cos\alpha = \frac{A_x}{A}$$

$$m = \cos\beta = \frac{A_y}{A}$$

$$n = \cos\gamma = \frac{A_z}{A}$$

Note that

$$l^2 + m^2 + n^2 = \cos^2\alpha + \cos^2\beta + \cos^2\gamma$$

$$= \left(\frac{A_x}{A}\right)^2 + \left(\frac{A_y}{A}\right)^2 + \left(\frac{A_z}{A}\right)^2$$

Since, from Section A.6, $A^2 = A_x^2 + A_y^2 + A_z^2$, it follows that $l^2 + m^2 + n^2 = 1$.

A.8 Example A1

Given the vectors $\mathbf{A} = \mathbf{i} - 2\mathbf{j} + 4\mathbf{k}$ and $\mathbf{B} = 3\mathbf{i} + \mathbf{j} - 2\mathbf{k}$, find $\mathbf{R} = \mathbf{A} + \mathbf{B}$.

Answer 1: The components of the given vectors are $A_x = 1$, $A_y = -2$, $A_z = 4$, $B_x = 3$, $B_y = 1$, $B_z = -2$. Thus, $R_x = A_x + B_x = 4$, $R_y = A_y + B_y = -1$, $R_z = A_z + B_z = 2$, so that

$$\mathbf{R} = \mathbf{A} + \mathbf{B} = 4\mathbf{i} - \mathbf{j} + 2\mathbf{k}$$

Answer 2: In Matlab, the vector **A** can be written as the following row matrix containing the components of **A**.

```
A = [1 -2 4]
```

Note that the components of a row vector are separated by spaces. Similarly, the vector **B** can be written as the following row matrix containing the components of **B**.

```
B = [3 1 -2]
```

The sum of these two vectors, **R**, can be found by writing **R** = **A** + **B** in Matlab as shown in Matlab Example A1a.

Matlab Example A1a

```
>> A = [1 -2 4]
A =
        1      -2       4

>> B = [3 1 -2]
B =
        3       1      -2

>> R = A + B
R =
        4      -1       2
```

Answer 3: In Matlab, the vector **A** can be written as the following column matrix containing the components of **A**.

```
A = [1; -2; 4]
```

Note that the components of a column vector are separated by semicolons. Similarly, the vector **B** can be written as the following column matrix containing the components of **B**.

```
B = [3; 1; -2]
```

The sum of these two vectors, **R**, can be found by writing **R** = **A** + **B** in Matlab as shown in Matlab Example A1b.

Matlab Example A1b

```
>> A = [1; -2; 4]
A =
        1
       -2
        4

>> B = [3; 1; -2]
B =
        3
        1
       -2

>> R = A + B
R =
        4
       -1
        2
```

A.9 Example A2

Find $A = |A|$ and $B = |B|$ for the vectors in Example A1.

Answer 1:

$$A = \sqrt{A_x^2 + A_y^2 + A_z^2} = \sqrt{1+4+16} = \sqrt{21} = 4.5826$$

$$B = \sqrt{B_x^2 + B_y^2 + B_z^2} = \sqrt{9+1+4} = \sqrt{14} = 3.7417$$

Answer 2: In Matlab, the magnitude of the vector **A** can be written as *norm*(**A**) as shown in Matlab Example A2.

Matlab Example A2

```
>> A = [1 -2 4]
A =
        1       -2        4
>> magA = norm(A)
magA =
     4.5826
>> B = [3 1 -2]
B =
        3        1       -2
>> magB = norm(B)
magB =
     3.7417
>>
```

A.10 Example A3

Find the unit vector λ_A in the direction of the vector $\mathbf{A}$ given in Example A1.

Answer 1:

$$\lambda_A = \frac{\mathbf{A}}{A} = \frac{1}{\sqrt{21}}\mathbf{i} - \frac{2}{\sqrt{21}}\mathbf{j} + \frac{4}{\sqrt{21}}\mathbf{k}$$

$$\lambda_A = 0.2182\mathbf{i} - 0.4364\mathbf{j} + 0.8729\mathbf{k}$$

Answer 2: In Matlab, the unit vector in the direction of $\mathbf{A}$ can be found as shown in Matlab Example A3.

Matlab Example A3

```
>> A = [1 -2 4]
A =
     1    -2     4

>> lambdaA = A/norm(A)
lambdaA =
    0.2182   -0.4364    0.8729
```

A.11 Example A4

Find the direction cosines of the vector $\mathbf{A}$ given in Example A1.

Answer 1:

$$l = \cos\alpha = \frac{A_x}{A} = \frac{1}{4.5826} = 0.2182$$

$$m = \cos\beta = \frac{A_y}{A} = \frac{-2}{4.5826} = -0.4364$$

$$n = \cos\gamma = \frac{A_z}{A} = \frac{4}{4.5826} = 0.8729$$

Answer 2: In Matlab, the direction cosines of $\mathbf{A}$ can be found as shown in Matlab Example A4.

Matlab Example A4

```
>> A = [1 -2 4]
A =
       1     -2      4

>> l = A(1)/norm(A)
l =
     0.2182

>> m = A(2)/norm(A)
m =
    -0.4364

>> n = A(3)/norm(A)
n =
     0.8729
```

Problems

Use Matlab to find the answers to the following problems.

A-1   Given the vectors $\mathbf{A} = 2\mathbf{i} + 6\mathbf{j} - 3\mathbf{k}$ and $\mathbf{B} = 3\mathbf{i} - 3\mathbf{j} + 2\mathbf{k}$, find

    (a) A and B (c) $3\mathbf{A} - 4\mathbf{B}$
    (b) $\mathbf{A} + \mathbf{B}$ (d) $|\mathbf{A} - \mathbf{B}|$

A-2   Repeat Problem A-1 for $\mathbf{A} = 5\mathbf{i} + 2\mathbf{j} - 7\mathbf{k}$ and $\mathbf{B} = -2\mathbf{i} - 3\mathbf{j} + 4\mathbf{k}$

A-3   Given the vectors $\mathbf{A} = 2\mathbf{i} + 3\mathbf{j} - \mathbf{k}$, $\mathbf{B} = 4\mathbf{i} - 3\mathbf{j} + 2\mathbf{k}$, and $\mathbf{C} = \mathbf{i} + 2\mathbf{j} - 3\mathbf{k}$, find

    (a) $\mathbf{A} + \mathbf{B} + \mathbf{C}$ (c) $|\mathbf{A}|$
    (b) $\mathbf{A} + \mathbf{B} - \mathbf{C}$ (d) $|\mathbf{A} + \mathbf{B} + \mathbf{C}|$

A-4   Find the direction cosines and the direction angles α, β, and γ of the vector $\mathbf{A} = 2\mathbf{i} + 5\mathbf{j} - 3\mathbf{k}$.

A-5   Repeat Problem A-4 for $\mathbf{A} = 6\mathbf{i} - 5\mathbf{k}$.

A-6   Find the unit vector λ_A in the direction of the vector $\mathbf{A} = 5\mathbf{i} - 5\mathbf{j} + 10\mathbf{k}$. Express λ_A in terms of $\mathbf{i}$, $\mathbf{j}$, and $\mathbf{k}$.

Appendix B

The Scalar or Dot Product

The multiplication of a vector by a scalar was discussed in Appendix A. When we multiply a vector by another vector, we must define precisely what we mean. One type of vector product is called the *scalar* or *dot* product and is covered in this appendix. A second type of vector product is called the *vector* or *cross* product and is covered in Appendix C.

Prerequisite knowledge:
  Appendix A – Addition and Subtraction of Vectors

B.1 Definition of the Dot Product

The *scalar* or *dot* product and is written as $\mathbf{A} \bullet \mathbf{B}$ and read "A dot B". The dot product is defined by the relation

$$\mathbf{A} \bullet \mathbf{B} = AB\cos\phi \qquad\qquad (B.1)$$

where ϕ is the angle between $\mathbf{A}$ and $\mathbf{B}$. Since the dot product $AB\cos\phi$ has only a magnitude and not a direction, then $\mathbf{A} \bullet \mathbf{B}$ is a *scalar* quantity.

The dot product $\mathbf{A} \bullet \mathbf{B} = AB\cos\phi$ can be written as $\mathbf{A} \bullet \mathbf{B} = \left(A\cos\phi\right)B$ where $A\cos\phi$ is the magnitude of the projection of $\mathbf{A}$ on $\mathbf{B}$ as shown in Fig. B.1.

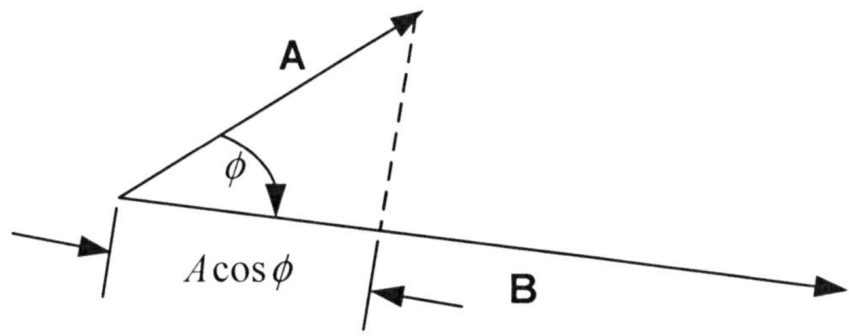

Figure B.1 $\mathbf{A} \bullet \mathbf{B} = \left(A\cos\phi\right)B$

The dot product can also be written as $\mathbf{A} \bullet \mathbf{B} = A(B\cos\phi)$ where $B\cos\phi$ is the magnitude of the projection of $\mathbf{B}$ on $\mathbf{A}$ as shown in Fig. B.2.

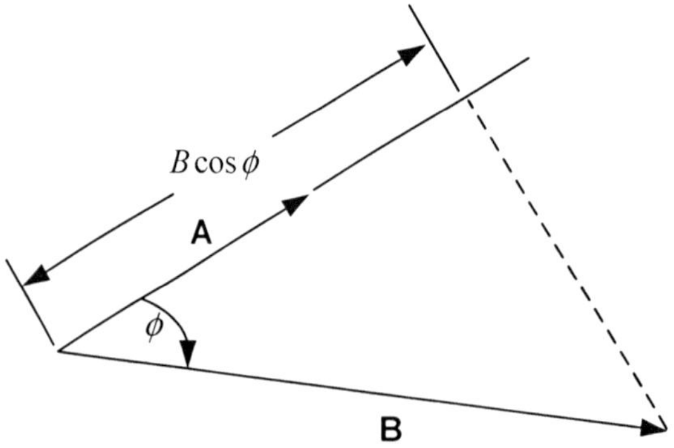

Figure B.2 $\mathbf{A} \bullet \mathbf{B} = A(B\cos\phi)$

From the definition of the dot product, $\mathbf{A} \bullet \mathbf{B} = AB\cos\phi$ and $\mathbf{B} \bullet \mathbf{A} = BA\cos\phi$. It is therefore clear that $\mathbf{A} \bullet \mathbf{B} = \mathbf{B} \bullet \mathbf{A}$ and the commutative law holds for the scalar product. The distributive law $\mathbf{A} \bullet (\mathbf{B}+\mathbf{C}) = \mathbf{A} \bullet \mathbf{B} + \mathbf{A} \bullet \mathbf{C}$ also holds and is illustrated for the special case shown in Fig. B.3 where $\mathbf{D} = \mathbf{B} + \mathbf{C}$.

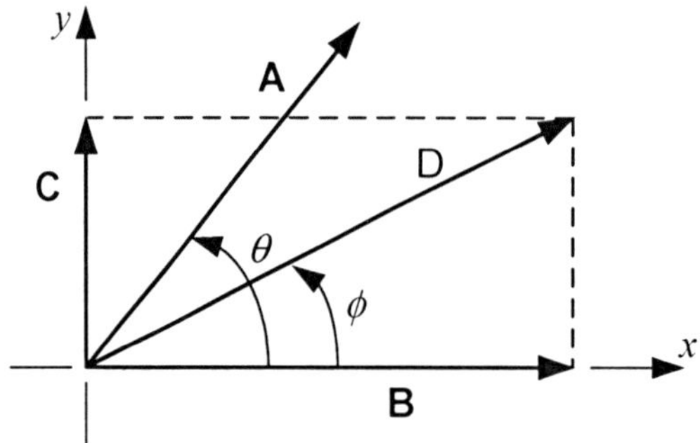

Figure B.3 $\mathbf{A} \bullet (\mathbf{B}+\mathbf{C}) = \mathbf{A} \bullet \mathbf{B} + \mathbf{A} \bullet \mathbf{C}$

Then

$$\mathbf{A} \bullet (\mathbf{B}+\mathbf{C}) = \mathbf{A} \bullet \mathbf{D}$$
$$= AD\cos(\theta - \phi)$$
$$= AD(\cos\theta\cos\phi + \sin\theta\sin\phi)$$

But, since $B = D\cos\phi$ and $C = D\sin\phi$,

$$\mathbf{A} \bullet (\mathbf{B}+\mathbf{C}) = AB\cos\theta + AC\sin\theta$$
$$= \mathbf{A} \bullet \mathbf{B} + \mathbf{A} \bullet \mathbf{C}$$

since the angle between $\mathbf{A}$ and $\mathbf{C}$ is $\dfrac{\pi}{2} - \theta$ and $\sin\theta = \cos\left(\dfrac{\pi}{2} - \theta\right)$.

B.2 Dot Product and Vector Components

The form of the dot product can be written conveniently in terms of its components in a rectangular coordinate system. Consider the two-dimensional case shown in Fig. B.4.

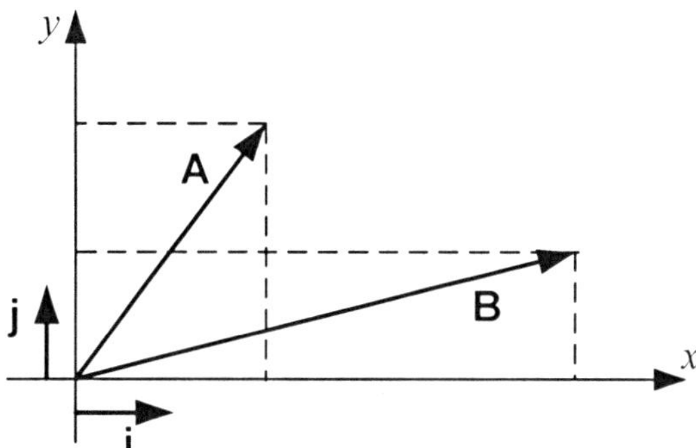

Figure B.4 Dot product in a rectangular coordinate system

The vectors **A** and **B** can be written in the component form $\mathbf{A} = A_x\mathbf{i} + A_y\mathbf{j}$ and $\mathbf{B} = B_x\mathbf{i} + B_y\mathbf{j}$. Then

$$\mathbf{A} \bullet \mathbf{B} = \left(A_x\mathbf{i} + A_y\mathbf{j} \right) \bullet \left(B_x\mathbf{i} + B_y\mathbf{j} \right)$$
$$= A_x B_x\, \mathbf{i} \bullet \mathbf{i} + A_y B_x\, \mathbf{j} \bullet \mathbf{i} + A_x B_y\, \mathbf{i} \bullet \mathbf{j} + A_y B_y\, \mathbf{j} \bullet \mathbf{j}$$

Since the unit vectors **i** and **j** are orthogonal (i.e., perpendicular), then from the definition of the scalar product $\mathbf{i} \bullet \mathbf{i} = \mathbf{j} \bullet \mathbf{j} = 1$ and $\mathbf{i} \bullet \mathbf{j} = \mathbf{j} \bullet \mathbf{i} = 0$. Thus, the scalar product can be written as

$$\mathbf{A} \bullet \mathbf{B} = A_x B_x + A_y B_y \tag{B.2}$$

Note that the dot product $\mathbf{A} \bullet \mathbf{B}$ must always involve the product of two *vectors*, and, since the result is a *scalar*, then an expression such as $\mathbf{A} \bullet \left(\mathbf{B} \bullet \mathbf{C} \right)$ has no meaning. On the other hand, the expression $\mathbf{A}\left(\mathbf{B} \bullet \mathbf{C} \right)$ does have a meaning.

B.3 Dot Product Properties

Consider the vector $\mathbf{A} = A_x\mathbf{i} + A_y\mathbf{j} + A_z\mathbf{k}$. From the orthogonality of the unit vectors described in Section B.2, it follows that

$$\mathbf{A} \cdot \mathbf{i} = A_x$$
$$\mathbf{A} \cdot \mathbf{j} = A_y$$
$$\mathbf{A} \cdot \mathbf{k} = A_z$$

Consider the definition of the dot product $\mathbf{A} \cdot \mathbf{B} = AB\cos\phi$. If $\mathbf{B} = \mathbf{A}$, then $\mathbf{A} \cdot \mathbf{A} = A^2$.

B.4 Example B1

Given the vectors $\mathbf{A} = \mathbf{i} - 2\mathbf{j} + 4\mathbf{k}$ and $\mathbf{B} = 3\mathbf{i} + \mathbf{j} - 2\mathbf{k}$ find $\mathbf{A} \cdot \mathbf{B}$.

Answer 1:

$$\begin{aligned}
\mathbf{A} \cdot \mathbf{B} &= A_x B_x + A_y B_y + A_z B_z \\
&= (1)(3) + (-2)(1) + (4)(-2) \\
&= 3 - 2 - 8 = -7
\end{aligned}$$

Answer 2: In Matlab, the dot product of vectors $\mathbf{A}$ and $\mathbf{B}$ can be written as *dot*(**A,B**) as shown in Matlab Example B1.

Matlab Example B1

```
>> A = [1 -2 4]
A =
     1    -2     4

>> B = [3 1 -2]
B =
     3     1    -2

>> AdotB = dot(A,B)
AdotB =
    -7
```

B.5 Example B2

Find the angle between the vectors $\mathbf{A}$ and $\mathbf{B}$ in Example B1.

Answer 1:

$$\mathbf{A} \cdot \mathbf{B} = AB\cos\phi = -7$$
$$A = \sqrt{21} \qquad B = \sqrt{14}$$
$$\cos\phi = \frac{\mathbf{A} \cdot \mathbf{B}}{AB} = \frac{-7}{\sqrt{21}\sqrt{14}} = -0.408$$
$$\phi = 114°$$

Answer 2: In Matlab, the solution can be found by writing the single Matlab equation shown in Matlab Example B2.

Matlab Example B2

```
>> A = [1 -2 4]
A =
     1    -2     4

>> B = [3 1 -2]
B =
     3     1    -2

>> phi = (acos(dot(A,B)/(norm(A)*norm(B))))*180/pi
phi =

   114.0948
```

Note carefully the need to use parentheses in the equation for *phi*. The Matlab function *acos* for the arc cosine gives the answer in radians. Thus, that result must be multiplied by $180/\pi$ to give the answer in degrees.

Problems

Where appropriate, use Matlab to find the answers to the following problems.

B-1   Given the vectors $\mathbf{A} = 3\mathbf{i} - 4\mathbf{j} - 2\mathbf{k}$ and $\mathbf{B} = 2\mathbf{i} + \mathbf{j} - 5\mathbf{k}$, find
    (a) $\mathbf{A} \bullet \mathbf{B}$ and $\mathbf{B} \bullet \mathbf{A}$.
    (b) the smaller angle between $\mathbf{A}$ and $\mathbf{B}$.
    (c) the component of $\mathbf{A}$ in the direction of $\mathbf{B}$?
    (d) the component of $\mathbf{B}$ in the direction of $\mathbf{A}$?

B-2   If $\mathbf{A} = 10\mathbf{i} + 5\mathbf{j} - 2\mathbf{k}$, determine A^2.

B-3   Given the vectors $\mathbf{A} = 3\mathbf{i} - 2\mathbf{j} + 5\mathbf{k}$ and $\mathbf{B} = 2\mathbf{i} + 8\mathbf{j} + 2\mathbf{k}$, show that $\mathbf{A}$ and $\mathbf{B}$ are perpendicular to each other.

B-4 $\mathbf{A} \bullet \mathbf{i} = 3$, $\mathbf{A} \bullet \mathbf{j} = 5$, and $\mathbf{A} \bullet \mathbf{k} = -2$.   Find $\mathbf{A}$.

B-5   If $\mathbf{A} = \mathbf{i} + 3\mathbf{j} - 2\mathbf{k}$ and $\mathbf{B} = 4\mathbf{i} - \mathbf{j} + 2\mathbf{k}$, find $(2\mathbf{A} + \mathbf{B}) \bullet (\mathbf{A} - 2\mathbf{B})$.

B-6   For what values of α are vectors $\mathbf{A} = \alpha\mathbf{i} - 2\mathbf{j} + \mathbf{k}$ and $\mathbf{B} = 2\alpha\mathbf{i} + \alpha\mathbf{j} - 4\mathbf{k}$ perpendicular?

Appendix C

The Vector or Cross Product

We saw in Appendix B that the dot product of two vectors is a scalar quantity that is a maximum when the two vectors are parallel and is zero if the two vectors are normal or perpendicular to each other. We now discuss another kind of vector multiplication called the *vector* or *cross product*, which is a vector quantity that is a maximum when the two vectors are normal to each other and is zero if they are parallel.

Prerequisite knowledge:
Appendix B – The Scalar or Dot Product

C.1 Definition of the Cross Product

The vector or cross product of two vectors is written as $\mathbf{A} \times \mathbf{B}$ and reads "A cross B." It is defined to be a third vector $\mathbf{C}$ such that $\mathbf{A} \times \mathbf{B} = \mathbf{C}$, where the magnitude of $\mathbf{C}$ is

$$C = |\mathbf{C}| = AB \sin \phi \tag{C.1}$$

and the direction of $\mathbf{C}$ is perpendicular to both $\mathbf{A}$ and $\mathbf{B}$ in a right-handed sense as shown in Fig. C.1. ϕ is the smaller angle between $\mathbf{A}$ and $\mathbf{B}$ and the direction of $\mathbf{C}$ is found by the following rule. Extend the fingers of your right hand along $\mathbf{A}$ and then curl them toward $\mathbf{B}$ as if you were rotating $\mathbf{A}$ through ϕ. Your thumb will then point in the direction of $\mathbf{C}$. The vector product $\mathbf{B} \times \mathbf{A}$ has a magnitude $BA \sin \phi$ but its direction, found by rotating $\mathbf{B}$ into $\mathbf{A}$ through ϕ, is opposite to that of $\mathbf{C}$. Therefore,

$$\mathbf{B} \times \mathbf{A} = -\mathbf{C} = -(\mathbf{A} \times \mathbf{B}) \tag{C.2}$$

and the commutative law does not hold for the cross product.

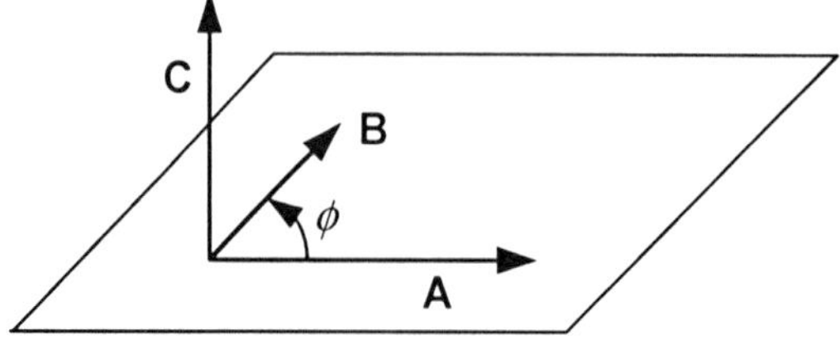

Figure C.1 $\mathbf{A} \times \mathbf{B} = \mathbf{C}$

C.2 Distributive Law for the Cross Product

The distributive law $\mathbf{A}\times(\mathbf{B}+\mathbf{C})=(\mathbf{A}\times\mathbf{B})+(\mathbf{A}\times\mathbf{C})$ holds in general for the cross product and is illustrated for the special case shown in Fig. C.2 where $\mathbf{A}$, $\mathbf{B}$, and $\mathbf{C}$ all lie in the x-y plane and $\mathbf{D} = \mathbf{B} + \mathbf{C}$. The $+\mathbf{k}$ direction is out of the paper, so from the right-hand rule, $\mathbf{A}\times\mathbf{D}$ is into the paper or in the $-\mathbf{k}$ direction. Similarly, $\mathbf{A}\times\mathbf{C}$ is in the $+\mathbf{k}$ direction and $\mathbf{A}\times\mathbf{B}$ is in the $-\mathbf{k}$ direction. We can then write

$$\mathbf{A}\times(\mathbf{B}+\mathbf{C})=\mathbf{A}\times\mathbf{D}$$
$$=-\mathbf{k}AD\sin(\theta-\phi)$$
$$=-\mathbf{k}AD\left(\sin\theta\cos\phi-\cos\theta\sin\phi\right)$$

But, since $B = D\cos\phi$ and $C = D\sin\phi$,

$$\mathbf{A}\times(\mathbf{B}+\mathbf{C})=-\mathbf{k}AB\sin\theta+\mathbf{k}AC\cos\theta$$
$$=(\mathbf{A}\times\mathbf{B})+(\mathbf{A}\times\mathbf{C})$$

since $\cos\theta=\sin\left(\dfrac{\pi}{2}-\theta\right)$ and the angle between $\mathbf{A}$ and $\mathbf{C}$ is $\left(\dfrac{\pi}{2}-\theta\right)$.

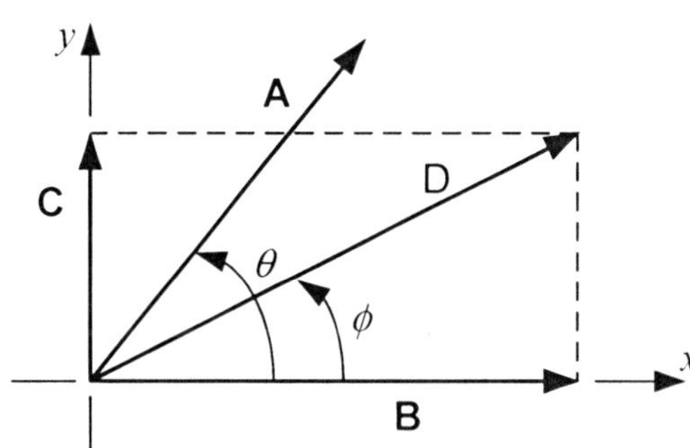

Figure C.2 $\mathbf{A}\times(\mathbf{B}+\mathbf{C})=(\mathbf{A}\times\mathbf{B})+(\mathbf{A}\times\mathbf{C})$

C.3 Cross Product and Vector Components

We now wish to find the components of $\mathbf{C}=\mathbf{A}\times\mathbf{B}$ in the rectangular coordinate system shown in Fig. C.3 if we know the components of $\mathbf{A}$ and $\mathbf{B}$. From the definition of the cross product, if two vectors are parallel, then $\phi=0$, $\sin\phi=0$, and their cross product is zero. In particular, the cross product of a vector with itself is always zero. Therefore, $\mathbf{i}\times\mathbf{i}=\mathbf{j}\times\mathbf{j}=\mathbf{k}\times\mathbf{k}=0$.

If two vectors are perpendicular, then $\phi=\pi/2$, $\sin\phi=1$, and the magnitude of their cross product is equal to the product of the magnitudes of the two vectors and the direction of the cross product is given by the right-hand rule. In particular, $\mathbf{i}\times\mathbf{j}$ is in the direction of **k** (rotate **i** into **j** with the fingers of your right hand and watch your thumb) and has a magnitude of unity. But this is just the unit vector **k**. Thus, in a similar way,

$$\mathbf{i}\times\mathbf{j} = \mathbf{k} \qquad \mathbf{j}\times\mathbf{i} = -\mathbf{k} \qquad \mathbf{k}\times\mathbf{i} = \mathbf{j} \qquad \mathbf{i}\times\mathbf{k} = -\mathbf{j} \qquad \mathbf{j}\times\mathbf{k} = \mathbf{i} \qquad \mathbf{k}\times\mathbf{j} = -\mathbf{i}$$

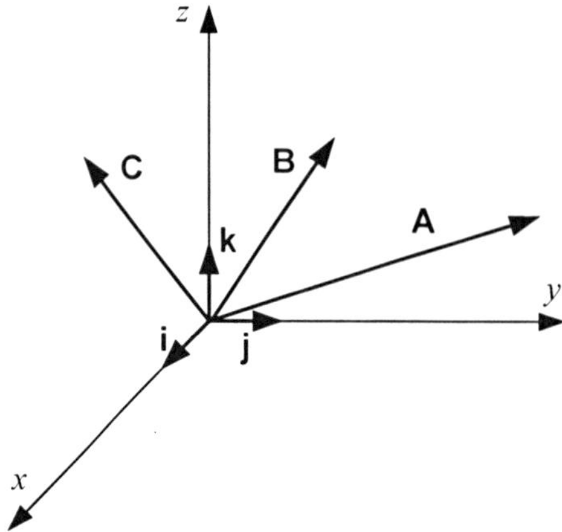

Figure C.3 $\mathbf{C} = \mathbf{A}\times\mathbf{B}$

Let us write the cross product

$$\mathbf{C} = \mathbf{A}\times\mathbf{B}$$
$$= \left(A_x\mathbf{i} + A_y\mathbf{j} + A_z\mathbf{k}\right)\times\left(B_x\mathbf{i} + B_y\mathbf{j} + B_z\mathbf{k}\right)$$
$$= \left(A_xB_x\mathbf{i}\times\mathbf{i}\right) + \left(A_xB_y\mathbf{i}\times\mathbf{j}\right) + \left(A_xB_z\mathbf{i}\times\mathbf{k}\right)$$
$$+ \left(A_yB_x\mathbf{j}\times\mathbf{i}\right) + \left(A_yB_y\mathbf{j}\times\mathbf{j}\right) + \left(A_yB_z\mathbf{j}\times\mathbf{k}\right)$$
$$+ \left(A_zB_x\mathbf{k}\times\mathbf{i}\right) + \left(A_zB_y\mathbf{k}\times\mathbf{j}\right) + \left(A_zB_z\mathbf{k}\times\mathbf{k}\right)$$

Using the above results to evaluate the cross products of the unit vectors, we can write

$$\mathbf{C} = \mathbf{A}\times\mathbf{B}$$
$$\mathbf{C} = \left(A_yB_z - A_zB_y\right)\mathbf{i} + \left(A_zB_x - A_xB_z\right)\mathbf{j} + \left(A_xB_y - A_yB_x\right)\mathbf{k}$$
$$= C_x\mathbf{i} + C_y\mathbf{j} + C_z\mathbf{k}$$

Therefore, if $\mathbf{C} = \mathbf{A}\times\mathbf{B}$, the components of **C** are given in terms of the components of **A** and **B** by

$$C_x = A_yB_z - A_zB_y$$
$$C_y = A_zB_x - A_xB_z$$
$$C_z = A_xB_y - A_yB_x$$

A useful way to remember the components of $\mathbf{C} = \mathbf{A}\times\mathbf{B}$ is to recognize that **C** can be written as the determinant

$$C = A \times B = \begin{vmatrix} \mathbf{i} & \mathbf{j} & \mathbf{k} \\ A_x & A_y & A_z \\ B_x & B_y & B_z \end{vmatrix} \qquad (C.3)$$

If you evaluate this determinant, you obtain the same result

$$\mathbf{C} = \left(A_y B_z - A_z B_y \right)\mathbf{i} + \left(A_z B_x - A_x B_z \right)\mathbf{j} + \left(A_x B_y - A_y B_x \right)\mathbf{k} \qquad (C.4)$$

C.4 Associative Law

The associative law does not in general hold for the vector product. Thus

$$\mathbf{A} \times (\mathbf{B} \times \mathbf{C}) \neq (\mathbf{A} \times \mathbf{B}) \times \mathbf{C}$$

as can be seen by the simple example shown in Fig. C.4. Since $\mathbf{A}$ and $\mathbf{B}$ are parallel, $(\mathbf{A} \times \mathbf{B}) \times \mathbf{C} = 0$. $(\mathbf{B} \times \mathbf{C})$ is a vector directed along the $+z$ axis (out of the paper), however, so that $\mathbf{A} \times (\mathbf{B} \times \mathbf{C})$ is a nonzero vector directed along the -y axis.

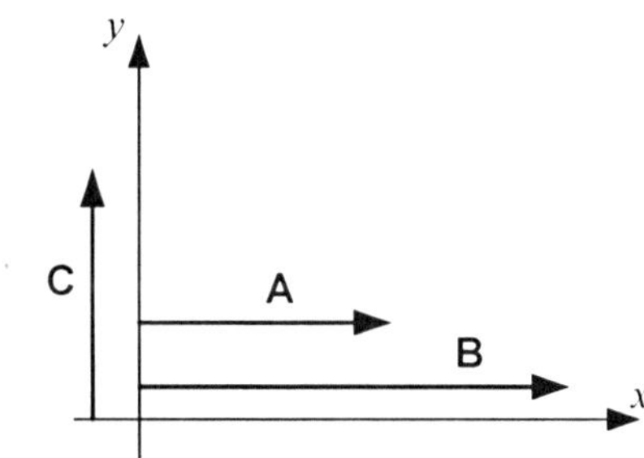

Figure C.4 $\mathbf{A} \times (\mathbf{B} \times \mathbf{C}) \neq (\mathbf{A} \times \mathbf{B}) \times \mathbf{C}$

C.5 Cross Product Geometric Properties

Consider the parallelogram with its sides formed by the vectors $\mathbf{A}$ and $\mathbf{B}$ as shown in Fig. C.5. The area of the parallelogram is $h|\mathbf{A}|$ where $h = |\mathbf{B}|\sin\phi$. Recall that the magnitude of the cross product is given by

$$|\mathbf{A} \times \mathbf{B}| = |A||B|\sin\phi$$

Therefore, in terms of $\mathbf{A}$ and $\mathbf{B}$, the area of the parallelogram is given by $|\mathbf{A} \times \mathbf{B}|$.

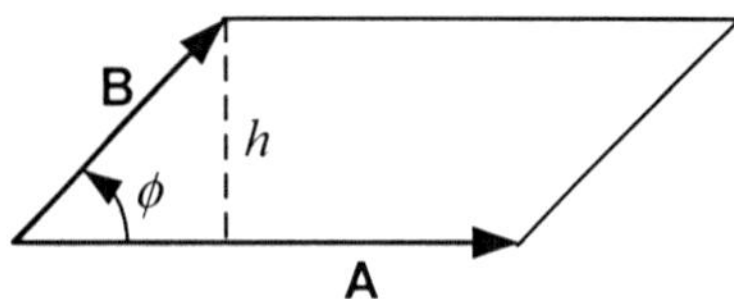

Figure C.5 The area of the parallelogram is $|\mathbf{A}\times\mathbf{B}|$

Consider the parallelepiped with its sides formed by the vectors **A**, **B**, and **C** as shown in Fig. C.6. The volume of the parallelepiped is (area of parallelogram formed by **A** and **B**) (height h) $=\left(|\mathbf{A}\times\mathbf{B}|\right)(\mathbf{C}\bullet\mathbf{n})$, where **n** is a unit vector parallel to $\mathbf{A}\times\mathbf{B}$. Since $\mathbf{A}\times\mathbf{B}=|\mathbf{A}\times\mathbf{B}|\mathbf{n}$, then in terms of **A**, **B**, and **C** the volume of the parallelepiped is given by $(\mathbf{A}\times\mathbf{B})\bullet\mathbf{C}$.

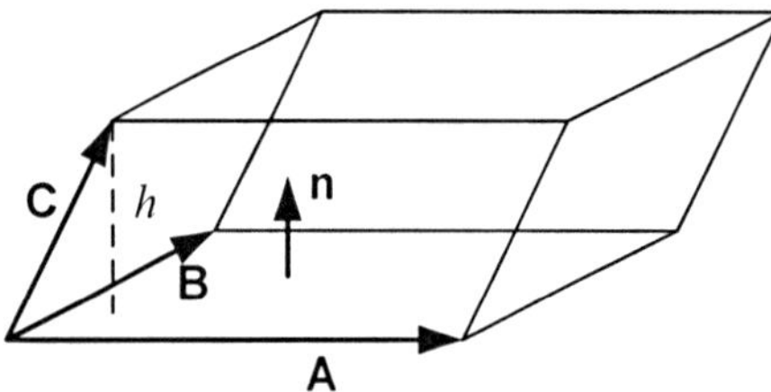

Figure C.6 The area of the parallelepiped is $(\mathbf{A}\times\mathbf{B})\bullet\mathbf{C}$

C.6 Scalar Triple Product

The expression $\mathbf{D}\bullet(\mathbf{A}\times\mathbf{B})$ is called the *scalar triple product*. It can be evaluated in terms of the rectangular components of **A**, **B**, and **D** by letting $\mathbf{C}=\mathbf{A}\times\mathbf{B}$. Then

$$\mathbf{D}\bullet(\mathbf{A}\times\mathbf{B})=\mathbf{D}\bullet\mathbf{C}$$

$$\mathbf{D}\bullet(\mathbf{A}\times\mathbf{B})=D_x\left(A_yB_z-A_zB_y\right)+D_y\left(A_zB_x-A_xB_z\right)+D_z\left(A_xB_y-A_yB_x\right)$$

Note that the scalar triple product can be written as the determinant

$$\mathbf{D}\bullet\mathbf{A}\times\mathbf{B}=\begin{vmatrix} D_x & D_y & D_z \\ A_x & A_y & A_z \\ B_x & B_y & B_z \end{vmatrix} \tag{C.5}$$

C.7 Example C1

Given the vectors $\mathbf{A} = \mathbf{i} - 2\mathbf{j} + 4\mathbf{k}$ and $\mathbf{B} = 3\mathbf{i} + \mathbf{j} - 2\mathbf{k}$, find $\mathbf{A} \times \mathbf{B}$ and $\left| \mathbf{A} \times \mathbf{B} \right|$.

Answer 1:

$$\mathbf{A} \times \mathbf{B} = \begin{vmatrix} \mathbf{i} & \mathbf{j} & \mathbf{k} \\ 1 & -2 & 4 \\ 3 & 1 & -2 \end{vmatrix}$$

$$= \mathbf{i}(4-4) + \mathbf{j}(12+2) + \mathbf{k}(1+6)$$

$$= 14\mathbf{j} + 7\mathbf{k}$$

$$\left| \mathbf{A} \times \mathbf{B} \right| = \sqrt{196+49} = \sqrt{245} = 15.7$$

Answer 2: In Matlab, the cross product of vectors $\mathbf{A}$ and $\mathbf{B}$ can be written as *cross*(**A**,**B**) as shown in Matlab Example C1.

Matlab Example C1

```
>> A = [1 -2 4]
A =
      1    -2     4

>> B = [3 1 -2]
B =
      3     1    -2

>> C = cross(A,B)
C =
      0    14     7

>> magC = norm(C)
magC =

   15.6525
```

C.8 Example C2

Use $|\mathbf{A}\times\mathbf{B}| = AB\sin\phi$ to find the angle ϕ between $\mathbf{A}$ and $\mathbf{B}$ in Example C1, and compare with the result of Example B2.

Answer 1:

$$\sin\phi = \frac{\mathbf{A}\times\mathbf{B}}{AB}$$

$$= \frac{\sqrt{245}}{\sqrt{21}\sqrt{14}} = 0.915$$

$$\phi = 180^\circ - 66^\circ = 114^\circ$$

Answer 2: In Matlab the solution can be found by writing the single Matlab equation shown in Matlab Example C2.

Matlab Example C2

```
>> A = [1 -2 4]
A =

     1    -2     4

>> B = [3 1 -2]
B =

     3     1    -2

>> phi = 180 - (asin(norm(cross(A,B))/(norm(A)*norm(B))))*180/pi
phi =

  114.0948
```

Note carefully the need to use parentheses in the equation for *phi*. The Matlab function *asin* for the arc sine gives the answer in radians. Thus, that result must be multiplied by $180/\pi$ to give the answer in degrees. Also note that the angle *phi* is greater than 90 degrees (as can be determined by plotting the vectors) and therefore the arc sine result must be subtracted from 180 degrees.

Problems

Where appropriate, use Matlab to find the answers to the following problems.

C-1 If $\mathbf{A} = 5\mathbf{i} - \mathbf{j} - 2\mathbf{k}$ and $\mathbf{B} = 2\mathbf{i} + 3\mathbf{j} - \mathbf{k}$, find
   (a) $\mathbf{A} \times \mathbf{B}$ and $\mathbf{B} \times \mathbf{A}$
   (b) $|\mathbf{A} \times \mathbf{B}|$
   (c) $\sin \phi$ and ϕ where ϕ is the smaller angle between $\mathbf{A}$ and $\mathbf{B}$.

C-2 If $\mathbf{A} = 5\mathbf{i} - \mathbf{j} - 2\mathbf{k}$ and $\mathbf{B} = 2\mathbf{i} + 3\mathbf{j} - \mathbf{k}$, find $\mathbf{A} \times \mathbf{B}$, $(\mathbf{A} \times \mathbf{B}) \bullet \mathbf{B}$, and $(\mathbf{A} \times \mathbf{B}) \bullet \mathbf{A}$.

C-3 If $\mathbf{A} = 3\mathbf{i} - 2\mathbf{j} + 4\mathbf{k}$, $\mathbf{B} = 2\mathbf{i} - 4\mathbf{j} + 5\mathbf{k}$, and $\mathbf{C} = \mathbf{i} + \mathbf{j} - 2\mathbf{k}$, find
   (a) $\mathbf{A} \times (\mathbf{B} \times \mathbf{C})$
   (b) $(\mathbf{A} \times \mathbf{B}) \times \mathbf{C}$

C-4 Evaluate
   (a) $2\mathbf{i} \times (3\mathbf{j} - 4\mathbf{k})$
   (b) $(\mathbf{i} + 2\mathbf{j}) \times \mathbf{k}$
   (c) $(2\mathbf{i} - 4\mathbf{j}) \times (\mathbf{i} + \mathbf{k})$

Appendix D

Matrices

We saw in Appendix A that the components of vectors could be written as row or column matrices. In general, matrices can have m rows and n columns. In this appendix we will see how matrices are used to solve simultaneous linear equations and we will explore the basic properties of matrices including the matrix transpose, multiplication of matrices, and the inverse of a matrix.

Prerequisite knowledge:
>Appendix C – The Vector or Cross Product

D.1 Solving Linear Equations

Suppose you want to solve the two simultaneous equations in two unknowns x_1 and x_2 shown in Eqs. (D.1) and (D.2). You could solve Eq. (D.2) for x_2 and substitute the result in Eq. (D.1) to find x_1, and then you could solve Eq. (D.1) for x_1 and substitute the result in Eq. (D.2) to find x_2. The results of doing this are shown in Eqs. (D.3) and (D.4).

$$a_{11}x_1 + a_{12}x_2 = c_1 \tag{D.1}$$

$$a_{21}x_1 + a_{22}x_2 = c_2 \tag{D.2}$$

$$x_1 = \frac{\left(c_1 a_{22} - c_2 a_{12}\right)}{\left(a_{11}a_{22} - a_{12}a_{21}\right)} \tag{D.3}$$

$$x_2 = \frac{\left(c_2 a_{11} - c_1 a_{21}\right)}{\left(a_{11}a_{22} - a_{12}a_{21}\right)} \tag{D.4}$$

Eqs. (D.3) and (D.4) have the same denominator that is called the *determinant* of a_{ij} and is written as

$$\Delta = \det\left(a_{ij}\right) = \begin{vmatrix} a_{11} & a_{12} \\ a_{21} & a_{22} \end{vmatrix} = \left(a_{11}a_{22} - a_{12}a_{21}\right) \tag{D.5}$$

Using this definition of determinant, we can rewrite Eqs. (D.3) and (D.4) as

$$x_1 = \frac{\begin{vmatrix} c_1 & a_{12} \\ c_2 & a_{22} \end{vmatrix}}{\Delta} = \frac{1}{\Delta}\left(c_1 a_{22} - c_2 a_{12}\right) \tag{D.6}$$

$$x_2 = \frac{\begin{vmatrix} a_{11} & c_1 \\ a_{21} & c_2 \end{vmatrix}}{\Delta} = \frac{1}{\Delta}\left(c_2 a_{11} - c_1 a_{21}\right) \tag{D.7}$$

Eqs. (D.6) and (D.7) is an example of using *Cramer's rule* to solve linear simultaneous equations. If we apply this rule to the three simultaneous equations shown in (D.8), we obtain the three solutions shown in Eqs. (D.9) – (D.11).

$$a_{11}x_1 + a_{12}x_2 + a_{13}x_3 = c_1$$
$$a_{21}x_1 + a_{22}x_2 + a_{23}x_3 = c_2 \tag{D.8}$$
$$a_{31}x_1 + a_{32}x_2 + a_{33}x_3 = c_3$$

$$x_1 = \frac{1}{\Delta}\begin{vmatrix} c_1 & a_{12} & a_{13} \\ c_2 & a_{22} & a_{23} \\ c_3 & a_{32} & a_{33} \end{vmatrix} \tag{D.9}$$

$$x_2 = \frac{1}{\Delta}\begin{vmatrix} a_{11} & c_1 & a_{13} \\ a_{21} & c_2 & a_{23} \\ a_{31} & c_3 & a_{33} \end{vmatrix} \tag{D.10}$$

$$x_3 = \frac{1}{\Delta}\begin{vmatrix} a_{11} & a_{12} & c_1 \\ a_{21} & a_{22} & c_2 \\ a_{31} & a_{32} & c_3 \end{vmatrix} \tag{D.11}$$

To evaluate the third-order determinant

$$\Delta = \begin{vmatrix} a_{11} & a_{12} & a_{13} \\ a_{21} & a_{22} & a_{23} \\ a_{31} & a_{32} & a_{33} \end{vmatrix} \tag{D.12}$$

we define the *minor* of a given element as that determinant formed by the remaining elements after the row and column containing the given element have been removed. Thus, if μ_{11} is the minor of a_{11}, then

$$\mu_{11} = \begin{vmatrix} a_{22} & a_{23} \\ a_{32} & a_{33} \end{vmatrix} \tag{D.13}$$

The value of the third-order determinant Δ in Eq. (D.12) is then

$$\Delta = \begin{vmatrix} a_{11} & a_{12} & a_{13} \\ a_{21} & a_{22} & a_{23} \\ a_{31} & a_{32} & a_{33} \end{vmatrix} = a_{11}\mu_{11} - a_{12}\mu_{12} + a_{13}\mu_{13}$$

$$= a_{11} \begin{vmatrix} a_{22} & a_{23} \\ a_{32} & a_{33} \end{vmatrix} - a_{12} \begin{vmatrix} a_{21} & a_{23} \\ a_{31} & a_{33} \end{vmatrix} + a_{13} \begin{vmatrix} a_{21} & a_{22} \\ a_{31} & a_{32} \end{vmatrix} \tag{D.14}$$

$$= a_{11}a_{22}a_{33} - a_{11}a_{23}a_{32} - a_{12}a_{21}a_{33} + a_{12}a_{23}a_{31} + a_{13}a_{21}a_{32} - a_{13}a_{22}a_{31}$$

Matrices

We have just seen that it can be cumbersome to solve the three simultaneous equations

$$a_{11}x_1 + a_{12}x_2 + a_{13}x_3 = c_1 \tag{D.15}$$
$$a_{21}x_1 + a_{22}x_2 + a_{23}x_3 = c_2 \tag{D.16}$$
$$a_{31}x_1 + a_{32}x_2 + a_{33}x_3 = c_3 \tag{D.17}$$

Using matrices can simplify the solution of these equations. We define a 3 x 3 matrix **A** of the coefficients a_{ij} as

$$\mathbf{A} = \begin{bmatrix} a_{11} & a_{12} & a_{13} \\ a_{21} & a_{22} & a_{23} \\ a_{31} & a_{32} & a_{33} \end{bmatrix} \tag{D.18}$$

We also define the two column vectors (3 x 1 matrices) **x** and **c** as

$$\mathbf{x} = \begin{bmatrix} x_1 \\ x_2 \\ x_3 \end{bmatrix} \tag{D.19}$$

and

$$\mathbf{c} = \begin{bmatrix} c_1 \\ c_2 \\ c_3 \end{bmatrix} \tag{D.20}$$

Then the three simultaneous equations given by Eqs. (D.15) – (D.17) can be written as the single matrix equation

$$\begin{bmatrix} a_{11} & a_{12} & a_{13} \\ a_{21} & a_{22} & a_{23} \\ a_{31} & a_{32} & a_{33} \end{bmatrix} \begin{bmatrix} x_1 \\ x_2 \\ x_3 \end{bmatrix} = \begin{bmatrix} c_1 \\ c_2 \\ c_3 \end{bmatrix} \tag{D.21}$$

or in more compact form as

$$\mathbf{Ax} = \mathbf{c} \tag{D.22}$$

For Eq. (D.21) to be equivalent to Eqs. (D.15) – (D.17), we see that *matrix multiplication* involves multiplying rows into columns. Thus, the first equation in (D.21) is formed by multiplying a_{11} by x_1 and then adding a_{12} times x_2 and then adding a_{13} times x_3 and setting this sum equal to c_1. This gives Eq. (D.15). To get Eq. (D.16), just multiply the second row of (D.21) into the column vector $\mathbf{x}$ and equate it to c_1. Applying the same procedure to the third row in (D.21) will give Eq. (D.17).

It is important to note that, in Matlab, any time you use the multiplication symbol * it is going to do this *row into column* matrix multiplication. Consider the matrix multiplication

$$\mathbf{z} = \mathbf{My} \tag{D.23}$$

For matrix multiplication to work in Eq. (D.23), the number of columns of $\mathbf{M}$ must be equal to the number of rows of $\mathbf{y}$. The number of rows of $\mathbf{z}$ will be equal to the number of rows of $\mathbf{M}$ and the number of columns of $\mathbf{z}$ will be equal to the number of columns of $\mathbf{y}$.

The *transpose* of a matrix $\mathbf{A}$ is formed by exchanging rows and columns and is denoted by $\mathbf{A}'$. Thus, the transpose of the matrix in Eq. (D.18) is

$$\mathbf{A}' = \begin{bmatrix} a_{11} & a_{21} & a_{31} \\ a_{12} & a_{22} & a_{32} \\ a_{13} & a_{23} & a_{33} \end{bmatrix} \tag{D.24}$$

Taking the transpose of a row vector turns it into a column vector, and taking the transpose of a column vector turns it into a row vector. Matlab Example D1 shows some examples of matrix multiplication and transpose. Note that we can form the 2 x 2 matix

$$\mathbf{M} = \begin{bmatrix} 2 & 3 \\ 4 & 5 \end{bmatrix}$$

by writing the Matlab statement

```
M = [2 3;4 5]
```

The first example shows that

$$\mathbf{z} = \begin{bmatrix} 2 & 3 \\ 4 & 5 \end{bmatrix}\begin{bmatrix} 2 \\ 3 \end{bmatrix} = \begin{bmatrix} 13 \\ 23 \end{bmatrix} \tag{D.25}$$

The second example shows that, if we try to multiply

$$\mathbf{w} = \begin{bmatrix} 2 \\ 3 \end{bmatrix}\begin{bmatrix} 2 & 3 \\ 4 & 5 \end{bmatrix},$$

we get an error because the number of columns of the first matrix does not match the number of rows of the second matrix. In the third example, we fix this by making $\mathbf{y}$ a row vector and computing

$$\mathbf{w} = \begin{bmatrix} 2 & 3 \end{bmatrix}\begin{bmatrix} 2 & 3 \\ 4 & 5 \end{bmatrix} = \begin{bmatrix} 16 & 21 \end{bmatrix} \tag{D.26}$$

Note that this gives a different result from Eq. (D.25) and we see that matrix multiplication does *not* commute. That is, in general $\mathbf{AB}$ is not equal to $\mathbf{BA}$ when $\mathbf{A}$ and $\mathbf{B}$ are matrices. However, it is true that $(\mathbf{AB})' = \mathbf{B}'\mathbf{A}'$ as shown in the fourth example in Matlab Example D1 where we see that

$$\mathbf{v} = \begin{bmatrix} 2 & 3 \end{bmatrix}\begin{bmatrix} 2 & 4 \\ 3 & 5 \end{bmatrix} = \begin{bmatrix} 13 & 23 \end{bmatrix} \tag{D.27}$$

In the final example we see that

$$\mathbf{z} = \begin{bmatrix} 1 & 2 \\ 3 & 4 \\ 5 & 6 \end{bmatrix}\begin{bmatrix} 2 \\ 4 \end{bmatrix} = \begin{bmatrix} 10 \\ 22 \\ 34 \end{bmatrix}$$

Matlab Example D1

```
>> y = [2;3]
y =
     2
     3

>> M = [2 3;4 5]
M =
     2     3
     4     5

>> z = M*y
z =
    13
    23

>> w = y*M
??? Error using ==> mtimes
Inner matrix dimensions must agree.

>> yr = y'
yr =
     2     3

>> w = yr*M
w =
    16    21

>> v = y'*M'
v =
    13    23

>> A = [1 2; 3 4; 5 6]
A =
     1     2
     3     4
     5     6

>> B = [2 4]'
B =
     2
     4

>> C = A*B
C =
    10
    22
    34
```

Matrix Inverse

We return now to Eqs. (D.21) and (D.22) where we are trying to solve for the three unknown values of $\mathbf{x}$ in terms of the known coefficients $\mathbf{A}$ and the known constants $\mathbf{c}$. We have the matrix equation

$$\mathbf{Ax} = \mathbf{c} \tag{D.28}$$

and the question is how can we solve for $\mathbf{x}$. We would like to just divide $\mathbf{c}$ by $\mathbf{A}$ but what does this mean? We only know how to do matrix multiplication. We begin by defining the identity matrix as

$$\mathbf{I} = \begin{bmatrix} 1 & 0 & 0 \\ 0 & 1 & 0 \\ 0 & 0 & 1 \end{bmatrix} \tag{D.29}$$

It is clear by doing the matrix multiplication that

$$\mathbf{Ix} = \mathbf{x} \tag{D.30}$$

We will define the matrix inverse $\mathbf{A}^{-1}$ by the equation

$$\mathbf{A}^{-1}\mathbf{A} = \mathbf{I} \tag{D.31}$$

Thus, if we pre-multiply a matrix by its inverse, we get the identity matrix. This will only work if $\mathbf{A}$ is a square matrix. If we then pre-multiply Eq. (D.28) by $\mathbf{A}^{-1}$ and use Eqs. (D.31) and (D.30), we obtain

$$\mathbf{A}^{-1}\mathbf{Ax} = \mathbf{A}^{-1}\mathbf{c}$$

$$\mathbf{Ix} = \mathbf{A}^{-1}\mathbf{c}$$

$$\mathbf{x} = \mathbf{A}^{-1}\mathbf{c} \tag{D.32}$$

Therefore, if we know $\mathbf{A}^{-1}$, we can find $\mathbf{x}$ from Eq. (D.32). How can we find $\mathbf{A}^{-1}$? We found the solutions by Cramer's rule in Eqs. (D.9) – (D.11) to be

$$x_1 = \frac{1}{\Delta}\begin{vmatrix} c_1 & a_{12} & a_{13} \\ c_2 & a_{22} & a_{23} \\ c_3 & a_{32} & a_{33} \end{vmatrix} = \frac{1}{\Delta}\left(c_1\mu_{11} - c_2\mu_{21} + c_3\mu_{31}\right) \tag{D.33}$$

$$x_2 = \frac{1}{\Delta}\begin{vmatrix} a_{11} & c_1 & a_{13} \\ a_{21} & c_2 & a_{23} \\ a_{31} & c_3 & a_{33} \end{vmatrix} = \frac{1}{\Delta}\left(-c_1\mu_{12} + c_2\mu_{22} - c_3\mu_{32}\right) \tag{D.34}$$

$$x_3 = \frac{1}{\Delta}\begin{vmatrix} a_{11} & a_{12} & c_1 \\ a_{21} & a_{22} & c_2 \\ a_{31} & a_{32} & c_3 \end{vmatrix} = \frac{1}{\Delta}\left(c_1\mu_{13} - c_2\mu_{23} + c_3\mu_{33}\right) \tag{D.35}$$

Writing Eqs. (D.33) – (D.34) in matrix form, we obtain

$$\begin{bmatrix} x_1 \\ x_2 \\ x_3 \end{bmatrix} = \frac{1}{\Delta}\begin{bmatrix} \mu_{11} & -\mu_{21} & \mu_{31} \\ -\mu_{12} & \mu_{22} & -\mu_{32} \\ \mu_{13} & -\mu_{23} & \mu_{33} \end{bmatrix}\begin{bmatrix} c_1 \\ c_2 \\ c_3 \end{bmatrix} \tag{D.36}$$

Comparing Eqs. (D.36) and (D.32) we see that the matrix inverse $\mathbf{A}^{-1}$ is given by

$$\mathbf{A}^{-1} = \frac{1}{\Delta}\begin{bmatrix} \mu_{11} & -\mu_{21} & \mu_{31} \\ -\mu_{12} & \mu_{22} & -\mu_{32} \\ \mu_{13} & -\mu_{23} & \mu_{33} \end{bmatrix} \tag{D.37}$$

Thus, if we know the matrix $\mathbf{A}$, we can find its inverse $\mathbf{A}^{-1}$ by performing the following three steps.

1. Evaluate the determinant of $\mathbf{A}$, denoted by Δ, and multiply the new matrix (to be $\mathbf{A}^{-1}$) by $1/\Delta$.
2. Insert minus signs in a checkerboard array starting in the 12 and 21 positions.
3. For the various entries of $\mathbf{A}^{-1}$, insert the minors of the corresponding entries in the *transpose* of $\mathbf{A}$. Thus, for example, μ_{12} is entered in the position of a_{21}; μ_{31} in the position of a_{13}; etc.

Note that the evaluation of $\mathbf{A}^{-1}$ requires division by Δ, the determinant of $\mathbf{A}$. If $\Delta = 0$, then $\mathbf{A}$ has no inverse and the matrix $\mathbf{A}$ is said to be *singular.*

It is easy to solve linear simultaneous equations in Matlab. You could use the Matlab function *det(A)* that computes the determinant of a square matrix $\mathbf{A}$ to help compute the Cramer's rule solutions given in Eqs. (D.33) – (D.35). However, the Matlab function *inv(A)* will find the inverse of a square matrix $\mathbf{A}$ so that you can solve Eq. (D.32) by just writing the Matlab statement

```
x = inv(A)*c
```
$$\tag{D.38}$$

An even better way to solve Eq. (D.32) is to use the Matlab left divide operator \

and write

$$\mathtt{x\ =\ A\backslash c} \tag{D.39}$$

While Eq. (D.38) works only if $\mathbf{A}$ is a square matrix, Eq. (D.39) will even work if $\mathbf{A}$ is rectangular. It does this by computing what is called a pseudo-inverse. Consider the equation

$$\mathbf{Ax} = \mathbf{c} \tag{D.40}$$

where $\mathbf{A}$ has more rows that columns. This means there are more linear equations than we need to get a unique solution. For example, if $\mathbf{A}$ has 3 rows and 2 columns then any two of the 3 linear equations will give a solution for the 2-dimensional $\mathbf{x}$. In general, there will be three different solutions. Using Eq. (D.39) will give a single solution that will be a kind of average of the three.

To see what actually gets computed, if we pre-multiply Eq. (D.40) by the transpose of $\mathbf{A}$, we obtain

$$\mathbf{A}'\mathbf{Ax} = \mathbf{A}'\mathbf{c} \tag{D.41}$$

$\mathbf{A}'\mathbf{A}$ will now be a square matrix so we can take its inverse. If we then pre-multiply Eq. (D.41) by $\left(\mathbf{A}'\mathbf{A}\right)^{-1}$, we can write

$$\left(\mathbf{A}'\mathbf{A}\right)^{-1}\mathbf{A}'\mathbf{Ax} = \left(\mathbf{A}'\mathbf{A}\right)^{-1}\mathbf{A}'\mathbf{c}$$

$$\mathbf{Ix} = \left(\mathbf{A}'\mathbf{A}\right)^{-1}\mathbf{A}'\mathbf{c}$$

from which $\mathbf{x}$ can be computed by

$$\mathbf{x} = \left(\mathbf{A}'\mathbf{A}\right)^{-1}\mathbf{A}'\mathbf{c} \tag{D.42}$$

The expression $\left(\mathbf{A}'\mathbf{A}\right)^{-1}\mathbf{A}'$ is called the *pseudo-inverse* of $\mathbf{A}$.

D.2 Example D2

Solve the following simultaneous equations for x_1, x_2, and x_3.

$$\begin{aligned}
5x_1 - 3x_2 + 7x_3 &= 4 \\
2x_1 + 6x_2 - x_3 &= -8 \\
3x_1 + 9x_2 + 2x_3 &= 5
\end{aligned} \tag{D.43}$$

We can write the equations in matrix form as

$$\begin{bmatrix} 5 & -3 & 7 \\ 2 & 6 & -1 \\ 3 & 9 & 2 \end{bmatrix}\begin{bmatrix} x_1 \\ x_2 \\ x_3 \end{bmatrix} = \begin{bmatrix} 4 \\ -8 \\ 5 \end{bmatrix} \tag{D.44}$$

which is of the form

$$\mathbf{Ax = c} \tag{D.45}$$

Using both Eqs. (D.38) and (D.39) in Matlab Example D2, we see that the solution is

$$\mathbf{x} = \begin{bmatrix} -5.2619 \\ 1.2302 \\ 4.8571 \end{bmatrix} \tag{D.46}$$

Matlab Example D2

```
>> A = [5 -3 7; 2 6 -1; 3 9 2]
A =
     5    -3     7
     2     6    -1
     3     9     2

>> c = [4 -8 5]'
c =
     4
    -8
     5

>> x = inv(A)*c
x =
   -5.2619
    1.2302
    4.8571

>> x = A\c
x =
   -5.2619
    1.2302
    4.8571
```

D.3 Example D3

Solve the following simultaneous equations for x_1 and x_2.

$$\begin{aligned}
x_1 + 0x_2 &= 3 \\
0x_1 + x_2 &= 6 \\
x_1 - x_2 &= 0
\end{aligned} \tag{D.47}$$

We can write the equations in matrix form as

$$\begin{bmatrix} 1 & 0 \\ 0 & 1 \\ 1 & -1 \end{bmatrix} \begin{bmatrix} x_1 \\ x_2 \end{bmatrix} = \begin{bmatrix} 3 \\ 6 \\ 0 \end{bmatrix} \tag{D.48}$$

which is of the form
$$\mathbf{A}\mathbf{x} = \mathbf{c} \tag{D.49}$$

But in this case, the matrix $\mathbf{A}$ is not square and therefore has no inverse. However, we can use the left divide operator \ in Matlab to compute a pseudo-inverse and the result is given in Matlab Example D3 as

$$\mathbf{x} = \begin{bmatrix} 4 \\ 5 \end{bmatrix} \tag{D.50}$$

This solution is shown in Fig. D.3 where we have plotted the three lines corresponding to the three equations in (D.43). Note that the solution is a "best guess" average of the three solutions corresponding to the intersections of any two lines.

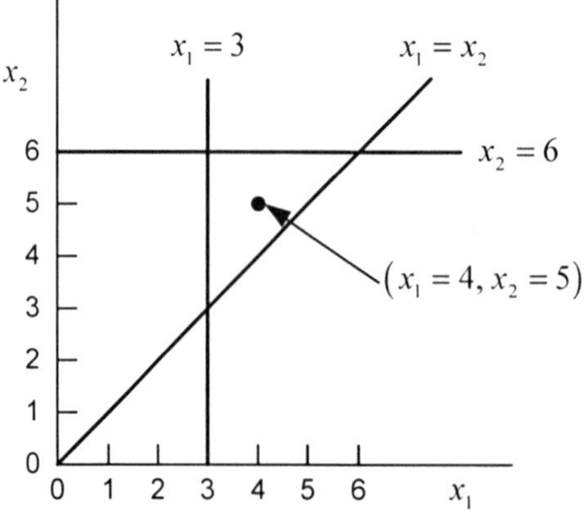

Figure D.1 Solution of Example D3

Matlab Example D3

```
>> A  =  [1  0;  0  1;  1  -1]
A  =
        1        0
        0        1
        1       -1

>> c  =  [3  6  0]'
c  =
        3
        6
        0

>> x  =  A\c
x  =
        4.0000
        5.0000
```

Problems

Use Matlab to solve the following problems.

D-1   Solve the following simultaneous equations for x_1, x_2, and x_3.

$$2x_1 - 7x_2 + 10x_3 = 4$$
$$-4x_1 + 2x_2 - 3x_3 = 12$$
$$x_1 + 6x_2 + 3x_3 = -5$$

D-2   Solve the following simultaneous equations for x_1, x_2, and x_3.

$$3x_1 - 2x_2 + 7x_3 = 8$$
$$9x_1 + 4x_2 - 7x_3 = 14$$
$$-1x_1 + 3x_2 - 2x_3 = -2$$

D-3   Use the Matlab left divide operator to solve the following three simultaneous equations for x_1 and x_2. Plot the three lines corresponding to the three equations and plot your solution.

$$2x_1 - x_2 = 2$$
$$x_1 + x_2 = 3$$
$$x_1 - x_2 = 0$$

Index